THE SOUTHERN UNIVERSE

LENNARD BICKEL

M

For my wife Pauline

First published 1975 by
THE MACMILLAN COMPANY OF AUSTRALIA PTY LTD
107 Moray Street, South Melbourne 3205
12 Berry Street, North Sydney 2060

Associated Companies in
London and Basingstoke, England
New York Dublin Johannesburg Madras

National Library of Australia
cataloguing in publication data

Bickel, Lennard
The Southern universe/ [by] Lennard Bickel. — South Melbourne, Vic.: Macmillan, 1975.
ISBN 0 333 17578 6.

1. Astronomy. I. Title.

520

Photoset by Pre Press, Melbourne
Printed in Hong Kong by Dai Nippon (HK) Ltd.

Title page photograph:

The Lagoon Nebula. This beautiful region of the great clouds behind Sagittarius (M.8, or NGC 2563) is a prime example of the rich fertility in the central regions of the Milky Way Galaxy. Here fresh young stars in open cluster burn brilliant against the gas-dust clouds where other young stars are forming. Here again are stages of creation where the supply of gas and dust is plentiful.

(Photo: Bart J. Bok — Arizona.)

AUTHOR'S NOTE AND ACKNOWLEDGMENTS

This book was written for people fascinated by the majesty of the night sky. It was planned as a journey across boundless space, to bring pleasure to the wondering mind and to give an inkling of the threshold at which modern astronomy stands.

The generosity and co-operation of many working astronomers made the book possible — and a delight to write. It is inevitable that some who helped may not be mentioned by name here, and that does not diminish my gratitude. However, my thanks must go to some of the leading figures in southern astronomy who opened doors — and often my mind — so this book could be developed. Professor Olin Eggen, Director of Mount Stromlo and Siding Spring Observatories, was a firm and major cornerstone; other pillars were Dr. Russell Cannon of the British Science Research Council team at Siding Spring, Dr. Robert Shobbrook of the Chatterton Astronomy Department, Sydney University, and the team at Mount Stromlo and Siding Spring . . . Drs. Alex Rodgers, Don Mathewson, Ken Freeman, Mike Bessell, Harry Hyland. Special mention is due to Vince Ford for producing, at Mount Stromlo, the colour photography of the true southern sky, published for the first time in this book, along with Dr. Mathewson's unique pictures of exploded stars in another "island" universe. The Director of the Anglo-Australian Telescope, Professor Joseph Wampler, and the Commissioning Astronomer, Professor S. C. B. ("Ben") Gascoigne — also of Mount Stromlo — have given valuable contributions, freely, while the resources of the photographic staff of the CSIRO Radio-Physics Division — especially John Masterton and Henry Armstrong — have been unstinted.

Appreciation is also due to Dr. Willson Hunter and Miss Lucy Letwinski of NASA, Canberra, as well as to American sources, in Washington, and at Mount Palomar, for their welcomed help. Debts — new and past — are also acknowledged to Dr. Alan Sandage, of California, to Sir Fred Hoyle (chairman of the Anglo-Australian Telescope Board), Sir Richard Woolley, former Astronomer-Royal and now directing South African astronomy, and to the irrepressible Professor Bart J. Bok — former Director of Mount Stromlo, now Professor Emeritus of Astronomy at the University of Arizona. Also, Professor Redman, of Cambridge, Professor Tom Gold, of Cornell, and John Bolton of the CSIRO, Parkes, have helped in this attempt to fashion an enjoyable venture into the Southern Universe.

Lennard Bickel,
Sydney. May, 1975.

FOREWORD

by Olin J. Eggen, Professor of Astronomy, Australian National University; Director, Mount Stromlo and Siding Spring Observatories: Founding Member, Anglo-Australian Telescope Board.

Nearly every type of object of interest and importance in the rapid expansion of our knowledge of the Universe is represented in our southern skies by its nearest, brightest, or most pathological example — the nearest galaxy, the closest star, the greatest splendour of the Milky Way. Australian astronomers have been leaders in the scientific exploitation of the southern heavens, and they are now being joined by their northern colleagues who have moved into the south with their telescopes. Four of the world's largest telescopes, including the Anglo-Australian Reflector, have been constructed in the south in the last decade. And, almost yearly in this decade, new astronomical results have forced a reorganisation of our basic concepts of the Universe, and each of these discoveries has seemed to reduce our own individual importance to that Universe.

Six hundred years ago Earth was considered to be the centre of our Universe; four hundred years ago the Sun was thought to be that centre and Earth was reduced to a mere satellite: fifty years ago the Sun was found to be far from the centre, and forty years ago the "Universe" was recognised to be only one galaxy in a far larger Universe containing countless galaxies. It is little wonder that man has sought to make contact with other intelligent life "somewhere out there in space" — if only to share this increasing loneliness and dwindling importance.

It should be noted, of course, that there is no such separate thing as Southern Hemisphere Astronomy — as in the same sense there is no separate science called Radio Astronomy, or Optical Astronomy since these are two approaches to the whole science of Astronomy. However, the accelerating renaissance that this science has undergone in the last decade would not have been possible if astronomers were blinded to either the radio or optical wavelengths. But, this development would, perhaps, only be slightly slowed if we were blocked from seeing one-half the sky; and, to my mind, there is little doubt that given their choice most astronomers would opt for keeping the southern skies available!

Knowledge of the construction and evolution of the Universe is the only product of astronomy, and, therefore, this knowledge should be given the widest circulation. Lennard Bickel has been a very active agent in this circulation for many years; the present book, THE SOUTHERN UNIVERSE, caps his efforts.

O.J. Eggen,
Cerro Tololo,
Chile.

CONTENTS

Author's Note and Acknowledgments 3

Foreword 4

Introduction 6

The Move to the south 8

Anatomy of the Sun 10

Neighbours of the Sun 22

The magnificent Milky Way 32

Heart of The Milky Way 52

Lovely companions to The Milky Way 59

A cosmos full of island universes 73

An invisible universe is unveiled 80

The cosmos is yielding its secrets 88

Glossary 94

FEATURES: Earth's magnetic blanket 15, Time and the stars 24, The Southern Cross 28, Light — The great revealer 42, Light as a yardstick 43, The strange theory of the "Black Holes" 65, Riches in the clouds 67, Giant on the mountain 70, Midget stars with astonishing power 86.

INTRODUCTION

Astronomy is in the throes of a revolution unparalleled in the long history of the science. An explosion of knowledge has flowed across the mid-decades of this remarkable 20th century, thrusting back the horizons of the universe to distances ten times greater than was ever dreamed. New "windows" thrown open by the leaps in technology and electronics have unveiled a cosmos of incredible power and beauty. Advanced communications developed in radio astronomy and the Space Age show a violent, changing universe; the scene now is far more awesome than any created in the fearful imaginations of primitive men — or the early civilisations.

Those early peoples saw the same glittering majesty in the night sky as does modern man; but, they had only fantasy to use to explain the inexplicable. They inhabited their night sky with mythical creatures — and called them gods — which they invested with punitive powers over the destinies of human beings; a paganism which still exists. Even the stupendous Milky Way earned its name from the fable of a god-child torn from its mother's breast so that the milk spilt from its mouth across the heavens. In these early conjecturings the Earth was always the focal point of the creation — everything else revolved about Earth. It was the centre of the universe — and woe to those who dared think differently. Reprisal extended to death at the stake.

The dark age of ignorance on the universe spanned too many centuries — not until Copernicus (16th century) then Brahe, Kepler, Galileo (17th century) did intellectual courage force back the curtains. Isaac Newton (1642-1727) in his *Principia* laid the base for modern astronomy with the laws which keep the planets in their orbits. The Earth was no longer the centre of the universe. Scientific man was on his way along a path which has brought him today close to the threshold of understanding the origin of the universe, its evolution and outline, its constituents.

The past few centuries produced knowledge which shattered old fables and fancies. Persecution of Copernicus and Galileo — and their supporters — could not stem the flood of dazzling truth finally to burst on human awareness in this present century. And, while the 1900's opened with some scientists still denying the existence of the atom, within three decades astronomy had confronted mankind with a new, thrilling universe, a matchless cosmos in which the teeming island galaxies — millions of Milky Ways — were flying away into dark space at speeds which increased unbelievably with distance.

This was the incredible *Expanding* Universe.

Yet — classic though the discovery was — it, too, was but a stepping-stone to the present plateau of knowledge. It was the opening to an epoch of discovery more impactive and revolutionary than any in history; for, now, man is closer than ever to unravelling the tale of how and when the great universe began.

An "invisible" universe stands revealed. Hidden secrets, formerly beyond detection by human biological senses, have unfolded to man's electronic technology. Radiation detectors — short-wave and radio — show us a cosmos far bigger, more violent, than was suspected. They also bring new clues on the origin of the universe and promise a new revelation; for, the formerly sedate science of astronomy, the tranquil observations in a beautiful universe untouched by time, is faced with stabbing reality. Astronomy is thrust into the vortex of a question that has spun in human minds since time began. It has been drawn into the search for the origin of life.

The teeming stars of the Milky Way hidden from our unaided eyes are seen in this telescopic colour view of the famous Southern Cross constellation. The five stars of the Cross lay athwart the picture, with the brightest star, Alpha Crucis in the bottom corner and Gamma Crucis — a brilliant single orange-red sun — is in the opposite top corner. Clearly seen is the star-forming cloud of dark gas known as the Coalsack, somewhat below the well-known "Jewel Box".

(Photo: Vince Ford, Mt. Stromlo)

This "invisible" universe has been compelled to yield the presence of enormous supply of life-building molecules along the Southern Milky Way.

Dark, compact, dusty clouds east of Southern Cross, around the heartland of the galaxy itself, send radio patterns that are fingerprints of an astonishing array of molecules we know build into amino-acids that form the proteins of organised life-forms. There is strong reason to predict such building-bricks-for-life may be proven in galaxies outside the Milky Way, the "home" galaxy in which we reside.

Not long ago scientists would have laughed from the rostrum any astronomer who lectured on the positive possibility of other civilisations dotted across the cosmos. Reports of clouds of life's essentials with the mass of 10-million suns are heard now in thrilled attention. Dozens of crucial molecules have been detected. Mankind is faced with the real prospect that he is probably *not* alone in the universe. The old comforting concept of the complex miracle of life being confined beneath the protective atmosphere of planet Earth is suddenly — almost brutally — attacked. Because, in these vast clouds the molecules precious to life are not trivia, not oddities — they are the prevailing common factor!

This strengthens the probability that evolution of life is a part of the scheme of things in the universe, since there must be a multitude of places with conditions like those on Earth where these molecules can form into bio-life of varied forms.

This present era, thus, is becoming an epoch in understanding, and in answering the age-old questions — what is creation? . . . how did all this begin? And at this classical period of the long struggle to *Know* — it is understandable that the focus of world astronomy — and its theoretical sister, cosmology — should swing to the richest and most promising panoramas of the whole sky — the unsurpassed Southern Universe.

The first major optical telescope built in the southern hemisphere — at Siding Spring Mountain, in eastern Australia — will be joined by others in Chile, and by one on a 14,000-feet high mountain on Hawaii.

The move to the south has been called a "gold rush" by leading astronomers. And it was forseen and predicted in 1961. Dr. Bart J. Bok (then Director, Mount Stromlo, later professor of astronomy at the University of Arizona) said, "In years to come special attention must be given to study of the Southern Milky Way. It holds the keys to solution of the most important problems . . ."

That time is now here. The full armament of modern astronomy is turning on to the Southern Universe, to view scenes of rich star-clouds and beautiful galaxies. With the planning of the big telescopes has come a steady and increasing flow of requests to southern observatories from astronomers in many parts of the world wishing to work in the promising south. The noted American explorer of deep-space, Dr. Allan Sandage, of Mount Palomar, has observed, "The South is now Nirvana to astronomers — there are so many wonderful things in the sky . . ."

What are those wonderful things? Why is the promise so high in the regions of the universe seen from southern latitudes? And where does southern astronomy stand in this classical and historical period of the long search for the truth of total creation?

The purpose of this book is to seek answers to those questions, to show the grandeur of the Southern Universe with the finest photography possible, and to reveal the exciting promise in the Southern Universe.

"We shall now look deeper into the Southern sky, collect a whole new mass of data, and might learn more about the universe than has ever before been possible."
Sir Fred Hoyle, Chairman
Anglo-Australian Telescope Board.

THE MOVE TO THE SOUTH

The expanses of the Southern Universe are so rich, in unique and striking examples of celestial objects, the choice of subjects for study seems endless. Overhead in the south is the heartland of the home galaxy of man, the Hub of the Milky Way, buried in the clustered starfields which form the stellar metropolis of the heavens.

Thereabouts are clusters and colonies of giant suns, blue-white-green, orange, and red stars and the vast glowing clouds of gas which are cradles for infant suns; there are the fascinating dark fragments of cloud which transmit across light years the radio signals which are fingerprints of molecule formation we now know to be vital to forming life, in the right conditions.

There are places where unstable stars have exploded in past times and left their expanding material glowing like radiant curtains; and on the outskirts of this great galaxy of the Milky Way, low across the southern sky, shine ancient globular clusters — the oldest stars in man's ken, beacons which tell of a time when the Milky Way was different. Then, beyond them, are shell after shell of teeming galaxies, trailing delicate arms of stars and gas, all running out to the far-off oblivion of the edge of the Southern Universe . . . to the point beyond which we shall never see.

Here is a treasurehouse for astronomy! Yet, a vast sweep of southern sky — the concourse seen only below 35° South — is virgin territory to advanced observation. This rich region, from where the Milky Way sweeps behind Sagittarius to the Southern Cross and the celestial south pole with its circling external galaxies, is only now being scanned in depth and detail. It is a part of the universe shut off forever from the north.

The optical telescope, Zelenchukskaya, in Russia — now the world's most massive — will never turn its 6-metre (236-inch) reflecting mirror on these important southern regions. Neither can the previously largest telescope — the Hale 200-inch on Mount Palomar, in California. These big northern telescopes can "see" immense stretches of southern sky, of course. But, the further down they depress their elevation, the deeper is the perturbation, the greater the distortion. They must look along those disturbed upper layers of the world's air, the strata of atmosphere which make stars twinkle and cause the rising Moon to come up golden yellow and twice the size — before it gets directly overhead.

This situation has led, at long last, to decisions to build fine telescopes south of the equator. The first major southern telescope to be commissioned is the 3.9-metre (150-inch) on Siding Spring Mountain, in northern New South Wales. A joint undertaking funded by the British and Australian governments — at a cost of $16-million — it was long overdue. The Southern Universe has long given many clues on its importance to world astronomy, but, in the north, these have, until recently, been received casually. Only now, in this last quarter of the 20th century, has the focus turned on southern riches. Only now is the Southern Universe being systematically photographed in a survey from which 12,000 time-lapse photographs will form the world's first comprehensive pictorial catalogue of the southern sky.

This survey is a co-operative venture in which astronomers working in Chile — where other large telescopes are in preparation and are planned — are taking part. In Australia, the pictorial catalogue will be used in conjunction with maps of the sky which Australian radio-astronomers have made on outstanding radio sources in the south.

When nights are dark in the southern world

This bright barred spiral galaxy is in a crowded corner of the Southern Sky, in the Fornax constellation. Known as NGC 1365, it burns with light that takes 50-million years to reach Earth. The two narrow arms reach out for 50,000 light years and hold dark dust lanes and bands of stars which glow to the ends of the bar.

(Photo: A.A.T.)

— away from the glow of city lights — the universe burns with a majestic splendour unmatched in the north.

The Milky Way is a gleaming arch running horizon to horizon. Overhead, along the plane of this great star system, is the crowded Hub of local creation, the centre of the star-clustered heaven which moved the minds of the first race of archaic men to settle in arid Australia . . . more than 40,000 years back in time. And astronomy is as old as that, in Australia. Those men of antiquity built their first cultures and rituals around the profound mystery of the glowing universe.

Those sensitive, creative men who first found Australia used the night sky as a playground for their imagination. Rich legends were woven about the creation of the universe. The brightest stars were given roles of people, lovers requited and jilted, ogres and heroes.

These legends were current, still part of culture when white men brought the first telescopes to the South Seas.

Captain James Cook was among the first. Astronomy brought him to the South Pacific. The purpose of his first journey of exploration was to set up a temporary observatory on Tahiti — to observe the transit of Venus across the face of the Sun. The passage of the planet across the solar face was measured with surprising accuracy; it gave astronomers a more refined measurement of Earth's distance from the solar centre, and thus astronomers regarded the discovery of Australia's east coast on that voyage as a secondary achievement.

Yet, it was the east coast landings and *not* the transit of Venus which brought settlement to Australia. Even so, when the First Fleet finally dropped anchor in Sydney Harbour in 1788, astronomy still had a role. Second-Lieutenant William Dawes had been trained to observe a great comet which the famous Halley predicted would be seen in the south — but never was.

Dawes made the first methodic observations from the south. He was followed by the first official observatory, erected in 1821 by the Governor, Sir Thomas Brisbane, at Parramatta, west of Sydney. The wooden building housed two simple telescopes, clocks and instruments and two trained observers . . . and they collated the first catalogue of southern stars — the Parramatta Catalogue.

Since then, observatories have risen across the land; Sydney Observatory — now an historic building — was built in 1858. And, though it did not begin its life until the mid-1920's, the Mount Stromlo Observatory, near Canberra, was destined to fill the major role in national astronomy. Siding Spring Mountain, home of the giant AAT instrument, is the up-country station for Mount Stromlo. And where a hundred years or so ago there were no more than half-a-dozen trained observers in the land, now the force runs to hundreds of world-ranking scientists, engineers, mathmeticians, computer-analysts, photographers, all committed to meeting the challenge of the Southern Universe.

The scale of their work stretches from the nearby Moon to the rim of the known universe, a vast range that defies easy assimilation; but, to come to familiarity with this scale, we shall, in this book, pace out the Sun's backyard to look at its family of planets and the space they occupy, before we take our first step out toward the neighbouring stars. From there, in this explanatory and pictorial journey, we shall move across the southern Milky Way and into that infinite universe to reach the point where our astronomers, by looking back into vast time, hope to predict the future.

ANATOMY OF THE SUN

"At rest in the middle of everything is the Sun. For in this most beautiful temple, who would place this lamp in another or better position than that from which it can light up the whole thing at the same time? Thus, indeed, as though seated on a royal throne, the Sun governs the family of planets revolving round it."

Nicholas Copernicus
(C. 1510 AD)

The Sun is a star. Familiar and overwhelmingly dominant in our sky, it is more important to us than any other object in the entire universe. Created 5,000-million years ago from a great swirling cloud of gas and dust, it gave birth to the Earth and other planets. It nurtured and maintains all life. All earthly existence depends entirely on its good behavior, on the continuing balance of the forces controlling its furious nuclear furnace.

Our Sun is a star, of a certain type, and so is an important clue to the processes which keep the cosmic clock ticking over. Understanding the anatomy of the Sun tells us how other stars shine across the universe; yet, until this present century, the source of its life-giving light and power was a mystery.

The Sun is a vast ball of gas with a diameter of, roughly, 1,400,000 kilometres; roughly, because its outer boundaries are of tenuous gas which cannot be accurately defined. It fills a volume 1-million times greater than our world and dominates our sky because it is so massive; in fact, 99.9 per cent of all the material in the Solar System rests in the Sun. All the planets and their satellites, the asteroids and rubble, the meteors and the tens of thousands of comets make up .01 per cent.

This Solar mass produces the gravity which, like an invisible string, holds Earth and its neighbours in orbit, and without which we should all go racing off into dark endless space. So — the Sun not only gives us the energy for life, it holds us in place so that we may enjoy it. As well, its powerful radiance lights up the Solar System, with an energy output equivalent to 380-million-million-million-million watts of electric current.

This outflow floods the higher layers of our atmosphere, gives us daylight and, by refraction, releases the blue light into man's sky; it shines on the dead face of the Moon, on Venus and Mars, and Jupiter . . . for, like Earth, no other planet sheds a light of its own making.

Where, then, does all this marvellous energy come from?

Like all stars — our Sun is a nuclear powerhouse.

Its glowing gas is in a state physicists know as *plasma*. Some 5,000-million years ago, when the great cloud from which the Solar System was born swirled into a tight-packed eddy, fierce compression excited the simple atoms of hydrogen into such agitation their structure of one proton and a single surrounding electron was disrupted. The gas became *ionized* — which is what happens to gas in neon tubes. The gas became the fourth state of matter — neither liquid, solid, nor gas; it became plasma, a sea of fractured atoms in which the homeless protons and electrons charge about seeking a proper resting place. It was from that plasma that the young Sun's atomic fire was first lit.

The recombining of those atomic particles is called *fusion*, and it provides the type of atomic reaction men dream of creating in powerhouses on Earth. The same force has been unleashed in H-bombs, and, if it can be contained and controlled, it would offer cheap, limitless access to electric power from nuclear reactors.

But — the Sun provides conditions man can never copy. Deep down in the fury at the centre of the Sun the nuclear furnace gulps four million tons of gas *every second of time* to keep the atomic pot boiling, to maintain a steady central temperature of 14-million degrees Centigrade!

In that fantastic inferno madly gyrating atomic particles smash and collide and then fuse together again — and again — until they become heavier and more stable atoms of helium. The hydrogen-helium cycle is a re-assembly of the stripped atoms. This process

The Face of the Sun. A view of the Sun as never seen before, this remarkable and historic photograph was taken by American astronauts of the fourth crew to man the orbiting space workshop — Skylab.

The Sun's polar regions are marked by diminution of the surface granulation and the fierce turbulence of the gas of the còrona reveals the strong magnetic forces at work in the vast ball of gas.

Even more striking in this picture taken in the light of ionized helium is the colossal flare looping out like a tenuous handle to the star we live by: NASA scientists claim this outburst to be among the most spectacular flares recorded by anyone. They say — "Spanning nearly 900,000 kilometres over the face of the Sun, the flare gives the impression of a twisted sheet of gas in the process of unwinding."

(Photo: NASA and US Naval Research Lab.)

releases fragments of matter which convert to radiant energy.

Imagine conditions in that great ball of plasma! The breakdown of atomic material unleashes a flood of energy, homeless frantic fragments of atoms, short-wave intense radiation, ultra-violet and gamma waves, x-rays and infrared, and the longer wavelengths of visible light and radio. From more than 400,000 kilometres down in the Sun this fierce flood of colossal energy thrusts upward — against the mighty compression, against the inward crushing force of gravity. The nuclear fusion reactions keep the Sun in balance. They prevent it from implosion — and keep it shining. The struggle upward for 400,000 kilometres is not the end of the journey! At that stage the outpouring energy meets the layer of the Sun we call the photosphere — which is a further 128,000 kilometres deep. After that comes the 10,000-kilometres-thick chromosphere. By now the struggling energy has shed much heat and is released through the Sun's tenuous corona — that head of fiery gas we see flaring during full eclipse by the Moon.

In all directions — radially — the energy flows to create the solar wind which blows through the planetary system . . . and which we did not know existed until the coming of the Space Age. The solar wind is the force which always keeps the tails of comets facing away from the Sun — no matter in which direction they travel.

Today, we can understand how the font of all our food and energy is in that furious nuclear furnace, how it is the prime agent of weather on Earth. This planet spins on its inclined axis, so some part of its surface is always turning from dark into the solar light — making it seem as though the Sun is moving through our sky, rising and setting.

In this way the tilted Earth constantly presents the faces of its seas and oceans to the Sun for warming. We know this raises moisture which creates rain-clouds, which slake the thirsting lands, to nourish plants and awaken seeds. We also know now that its light feeds the leaves, plants and trees and is converted into nutrients for growth, to swell the grain and the fruit. All this from the Sun's fusion furnace.

We understand too, that Earth's tilted axis (23½ degrees) means that for one half of the annual orbit of the Sun the southern hemisphere is closer to the solar surface, and so has its summer. For the other six months the northern world is closer to the Sun. Intersecting those periods, the median radiation creates the in-between seasons, spring and autumn. So fine is the balance of our life support in the Solar System.

Also, we now know the Sun has great storms — magnetic storms in which electromagnetic forces released from those nuclear reactions spiral up and out from the plasma to create dark pockmarks on the Sun's face . . . smudges we call sun-spots. They fling energy out to disrupt the upper layers of Earth's atmosphere we use for long-distance radio, and they cause flares which fling fiery gas from the Sun's face millions of kilometres out into space.

The frequency of these spiralling sun-spots has revealed an eleven-year cycle in the Sun which we do not yet understand. And there are also losses of material in the solar atomic reactions which our laws of physics have not yet explained.

Yet, one indisputable factor *has* emerged. We now recognise, a little sadly, that our Sun cannot last forever. Humanity must live with the knowledge that its solar creator will die. And the Sun is already middle-aged. Could it but go on gulping its gas at the present rate it would live on for another 50,000-million years. But, the dynamics of stellar evolution will not allow such a stay of execution. The advancing age of the Sun will make more and more demands on its reserves. The balance in that mighty core will swing steadily from hydrogen to helium and the fuel burn-up will quicken; its heat will increase.

Long before our star is 5-billion years older the helium-hydrogen balance will have swung so far the Sun will begin to swell into a red giant which will heat up the surrounding space. Its fight for life will bring down fire and brimstone on the inner planets — in true biblical style!

The Sun will expand out almost to the orbit of Mars. Mercury, Venus, and Earth, will all melt and drip into the solar mass. But long before then, all life will have ended on Earth. The oceans will have boiled away, the rocks melted to lava. Later, the tired, depleted Sun will shrink into a stage we know as "white dwarf", the last shining stage before final extinction makes our star a dark cinder in the teeming Milky Way galaxy.

The ending of the Sun will be a quiet demise. It will not flare out in sudden agony as some stars do when they explode. It will fade away leaving darkness where — 10-billion years earlier in a swirling gas-dust cloud — it was born with its family of planets . . . and so gave life to man.

The passing will make little change to the overall aspect of the Milky Way. Our great Sun — for all its power and mighty radiance — is really a minor star among 50-thousand-million orange-coloured stars burning across the galaxy, many of which may well have planets of their own.

Our sky is blue, but space is eternally black

Our sky is blue but space is eternally black. Our rotating planet causes the Sun to rise over the eastern horizon each morning, and at once a blue sky appears overhead.

This is normal, fully expected. It is also to be wondered at, since the reflected sunlight from the Moon or the planets is not blue. Why then does our sky run blue with the first kiss of the direct solar light?

Some 20 kilometres high in the atmosphere, at a point astronauts call the 'interface', spacecraft returning from orbit show friction from collision with atoms and molecules of air. Light beams from the Sun meet with the same collisions and the molecules about 20 kilometres high are just the right size to impede the passage of the shorter wave blue light in the solar 'rainbow' band of colours. Thus blue light waves are knocked helter-skelter across the upper layers of air and create our blue sky. The only place we know in the universe where this happens is on beautiful Earth.

Throughout space the sky is black, and astronauts report how our sky changes from blue to purple to violet with altitude. On the surface of the Moon (pictured) the sky is jet black in blazing sunshine . . . and stars dwindle to tiny needle points of unwinking light.

(NASA picture)

THE FAMILY OF PLANETS

The Sun created a family of nine planets. Each is a world of its own and — like all children — each a separate character. There are also two distinct divisions in the family. These are the inner planets (moving outward from the Sun) — Mercury, Venus, Earth, Mars — and the outer, generally very much larger worlds of Jupiter, Saturn, Uranus, Neptune and distant, lonely Pluto. All can be seen crossing the southern sky.

Between the orbits of Mars and giant Jupiter — largest of the family — is a region known as the asteroid belt which forms a dividing line separating the rocky inner planets from the gaseous bulks of Jupiter, Saturn and Uranus. One theory suggests the collection of lumps of rock which roll around this region was once a planet which exploded. Latest calculations show, however, that all the rubble in the asteroid belt would not make a decent-sized moon.

Though there is wide difference in the materials and density of the inner and outer planets (so far as is known) there is no doubt that all were formed from that same primordial cloud in which the Sun was born. Discussion arises among astronomers and geophysicists only as to *how* these worlds actually accumulated into spheres. A very recent concept tries to meet many arguments with a valid theory of how the gas cloud, shrinking into the infant Sun, threw off huge blobs of its material at stages of its contraction.

The concept says the outer layers first thrown off would be primitive atomic material of the original cloud; this would give the kind of gaseous, light elements which make up the larger outer planets. With a few million years of nuclear reactions in the young shrinking Sun, that original primordial material would have been changed by nuclear reactions — thus making for smaller, more dense planets.

Until exploration of the Solar System is more advanced we may not know if this is the answer. Unlike study of the stars, the planets are a one-off affair for man. The volume of space filled by the Solar System and the size of the planets themselves, are too small for us to see other systems round other stars. Thus we cannot make comparisons and judgements . . . immense though the distances between the planets seem to us on Earth. But, since the scale of this system is our first step into outer space, it is well worth looking at.

Our world holds middle place in the order of size, but is the third fastest of the planets in the motion round the Sun. The inner worlds move much faster along their orbits than do the big outer planets. This is because (see Profiles of the Planets) their speed has to defeat a greater tug of solar gravity. Thus, Mercury speeds round the Sun in only 88 days while Pluto lumbers along on an orbit which takes 248 Earth years to complete.

Because the orbits of the planets vary so widely in distance and speed, they appear to be erratic in our sky, and it was this which earned them their name — planet originally meant "wanderer" to early astronomers.

This distance between Pluto and Mercury seems meaningless to our minds if we use earthly measures. It runs into thousands of millions of kilometres and is better expressed in light-time. For instance, light which moves at a constant speed of almost 300,000 kilometres per second (186,000 miles) takes eight minutes to cross the distance of 150-million kilometres from the Sun to Earth. Light-time from the Moon (the most distant object yet reached by manned space-flight) is a mere 1.25-seconds. The light-time to Pluto is close to 4½ hours (depending where it is positioned in its elliptical orbit).

The American space agency (NASA) at this time has two space craft — Pioneer 10 and Pioneer 11 — beyond Jupiter and heading for the outer rim of the Solar System, marked by Pluto's orbit. They were launched in April 1972, and April 1973, and so are a year's travelling time apart. Beyond mighty Jupiter their speed has picked up by the influence of gravity of the larger planets and they cruise along — automated ships sending back tiny signals across space — at more than 45,000 kilometres an hour, the highest speed ever attained by a man-made object. For all that, while light from the Sun reaches Pluto in 4½ hours — the spacecraft will spend more than eleven years to cover the same distance.

Out of the Solar System the Pioneers will pass from human ken. Power sources will die away; there will be no signals to detect. For all we know they may travel on forever across the galaxy; but — supposing they *could* be directed towards the nearest neighbouring star.

Suppose the high rate of travel could be maintained on that journey . . . the first great inter-stellar mission! In that odyssey the ships would face a journey 8,000 times more distant than the trek to the orbit of Pluto. They could not reach their destination until A.D. 92,000. They would spend more than 90,000 years in reaching the nearest star to our Sun. So vast is space, so enormous is the region the Sun and its planet family occupies.

Earth's magnetic blanket!

Mankind lives at the bottom of an ocean of air. This belt of life-supporting atmosphere can be measured out to several hundred kilometres, but — three-quarters of its total weight of 5,000-million-million tonnes lies below the altitude of Mount Everest . . . about eight kilometres deep.

And the upper layers of this vital and precious air belt wrap our planet round with a magnetic blanket that shields and protects all organic life from the lethal short-wave radiation that pours from the Sun. These upper layers of air are known as the Ionosphere.

The ionosphere is a strata of air affected by Earth's own magnetic field — based on the heavy metallic core of the planet — and by the outer radiation fields known as the Van Allen Belts which form what is now known as Earth's magnetosphere. In these fields the most energetic short wave radiation and fast-moving atomic particles are snared, and those which do sneak through meet the barrier of the magnetised ionosphere which deflects them back into space.

As well, the ionosphere serves the world's long-distance radio communications. Since it reflects short wave radiation out into space, it also reflects short-wave radio signals back to Earth. Using this, man is able to broadcast beyond the horizon. Since radio waves travel with the same speed as light — as do all electromagnetic emissions — short wave radio signals can travel around the world in fractions of a second by bouncing back and forth from the surface and the underlying strata of the ionosphere — Earth's magnetic blanket.

PROFILES OF THE PLANETS

MERCURY: The first planet out from the Sun, Mercury's rocky face broils in direct solar heat rising above 600 degrees Centigrade. A small planet, with a diameter equal to the distance across Australia and a mass only half as much again as the Moon, it is the speediest of the solar family. Mercury's year lasts 88 Earth days. It hurtles along its orbit — 58-million kilometres out from the Sun's face — at a speed of more than 50-kilometres per second, necessary to defeat the immense pull of the Sun's gravity. Recent television images of the planet's face from the NASA Mariner series show a crater-pitted surface with signs of melting under an extremely thin atmosphere. Mercury only shows one face to the Sun (as our Moon does to Earth) which means it has to rotate on its axis once every 88 days — once in every solar circuit.

VENUS: Closest to Earth in size and proximity, Venus follows an orbit 108-million

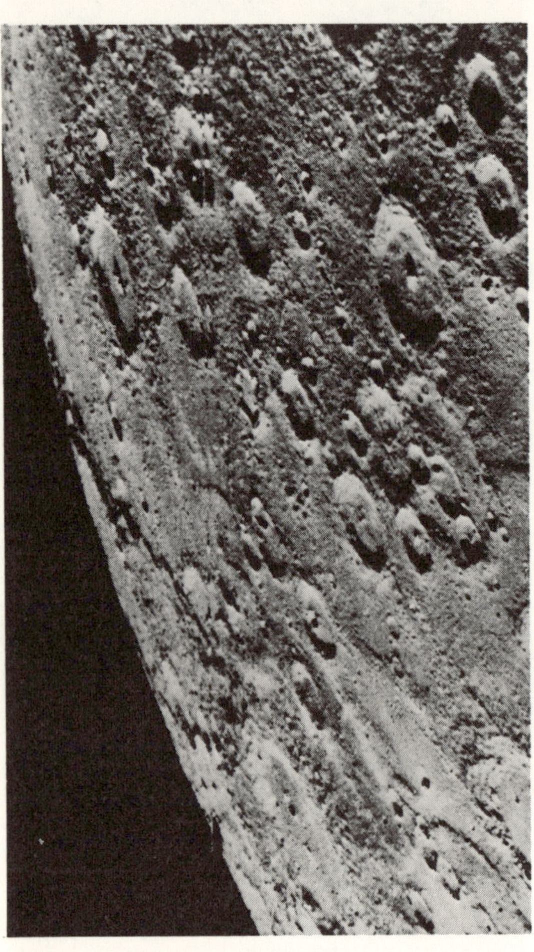

This view of Mercury's northern limb shows a prominent east-facing scarp extending from the limb near the middle of the photo southward for hundreds of kilometres. The linear dimension along the bottom of the photo is about 365 miles (580 kilometres). This picture was taken at a distance of 77,800 kilometres from Mercury. Mariner 10 passed the planet on the dark side (690 kilometres) from the surface. *(NASA Photo)*

Venus seen from Mariner Spacecraft. Earth's nearest neighbour in the Planet Family, and brightest of them all in our sky — other than our satellite Moon. The mystery is being stripped from the cloud-shrouded enigmatic Venus yet this picture shows an animated atmosphere of considerable beauty — even if some colour has been added by processing of the television image sent across 40-million kilometres to antennae on Earth. *(NASA photo)*

THE ORBITS OF THE PLANETS

SYMBOLS OF THE PLANETS

Symbols for each of the planets have been given since earliest times; the last one was accorded this century, when — in March 1930 — the faint light spot of distant Pluto was discovered. These are the symbols:

SUN — ☉

MERCURY — ☿

VENUS — ♀

EARTH — ⊕

MARS — ♂

JUPITER — ♃

SATURN — ♄

URANUS — ♅

NEPTUNE — ♆

PLUTO — ♇

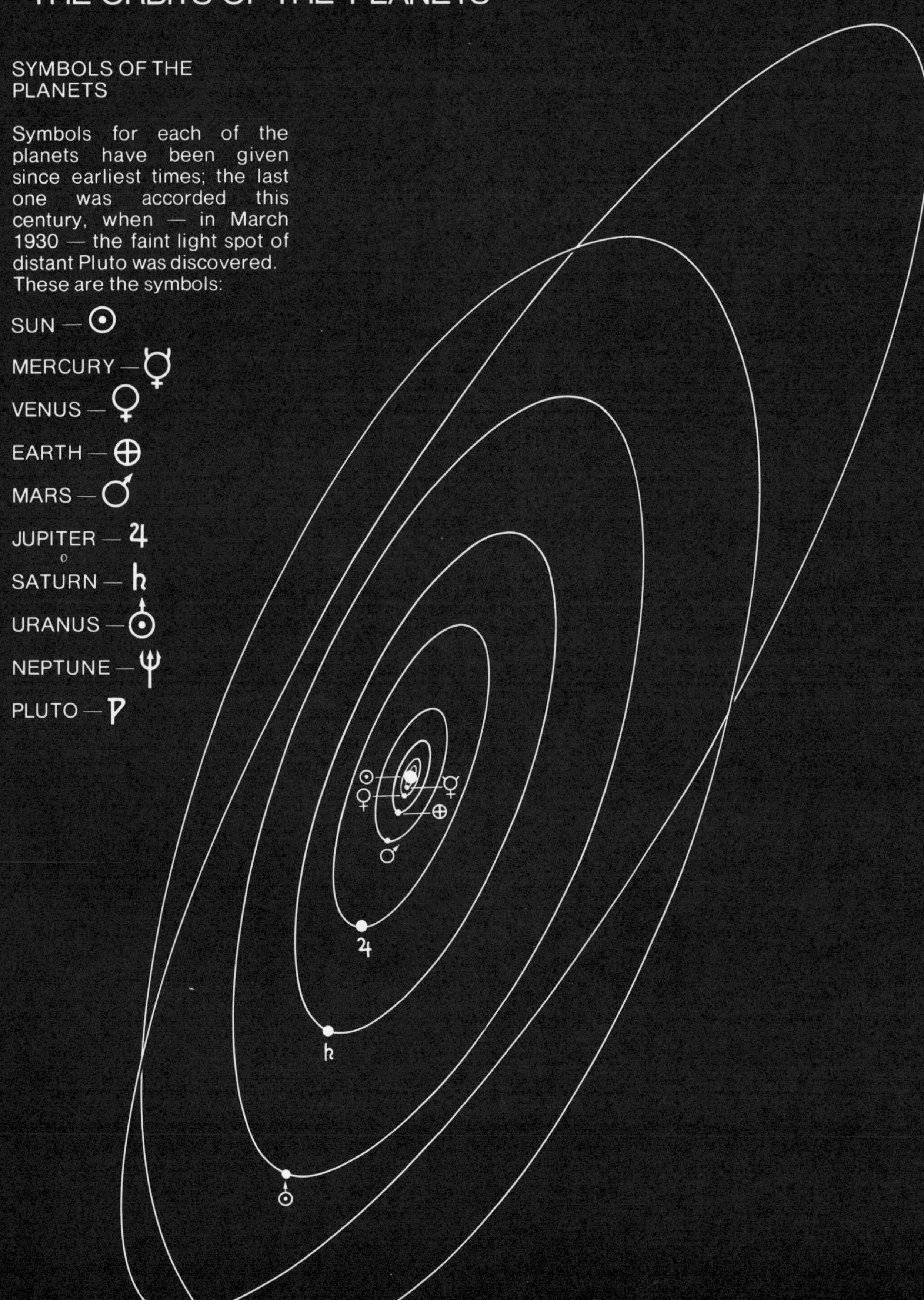

Earthrise on the Moon. No other picture in history showed so clearly how diminutive is our lovely planet, Earth, how thin is the atmosphere that shields our lives from harmful radiation.

Michael Collins took this picture in July, 1969, when Armstrong and Aldrin rose in their lunar landing-craft to rejoin the command ship in orbit about the Moon, thus ending the visit to the Sea of Tranquility, man's first step onto another world.

(N.A.S.A. Photo)

kilometres out from the Sun's face which brings it to within 38-million kilometres of the world's telescopes — 16-million kilometres nearer than the closest approach by Mars. Yet, no man has seen the face of Venus. Once held to be mysteriously beautiful, its dense white clouds mask conditions now known to be utterly hostile to man. Radar, and Russian and American automated probes, show a rough surface with temperatures above 500 degrees Centigrade and an atmospheric pressure about one hundred times greater than Earth's. Venus makes one turn about the Sun every 224 Earth-days at an orbital speed of 35 kilometres per second.

EARTH: Now judged to be 4,700-million years old, Earth is clearly the jewel among planets, the only world about the Sun which is tolerable to our life forms. The mean diameter of Earth is 12,742 kilometres and the annual parade in solar orbit takes 365.26 days at a mean solar distance of 149,500,000 kilometres. To achieve that the Earth moves at an orbital speed of 29.76 kilometres per second. Man knows more about Earth than any planet, but its inner structure and chemistry still hold many puzzles, even though tens of thousands of holes have been bored into the surface, some down as far as 10 kilometres. Man inhabits the only planet known to possess oceans and an atmosphere (capable of fine measurement out to more than 300 kilometres) which contains moisture and the life-supporting elements of oxygen and nitrogen in rich amounts. Earth's axis is tilted 23½ degrees against the plane of its orbit round the Sun and this gives rise to the alternate seasons between the north and southern climes of the world. The planet spins on its axis in the same general direction as it circles the Sun.

MARS: Long thought to be a sister planet to Earth and a possible abode of life, Mars has proved a disappointment since visiting spaceships of the last decade revealed details of its atmosphere and terrain. Mars circles the Sun at a mean distance of 228-million kilometres — 78-million kilometres further out than Earth, at a speed of 24 kilometres per second. It has a diameter a little more than half man's home planet — 6,664 kilometres. Thus, its mass is only one-tenth that of Earth while a Martian day lasts 24 hours, 37 minutes and its year 687 Earth-Days. Mars is now believed virtually devoid of surface water and plant life, and to have the most

Mars, long thought to be a possible abode of life, has proved a disappointment since spaceships of the last decade have revealed its nature as a rugged, barren and virtually waterless mass. Mars mass is only one-tenth that of Earth. The legendary "canals" of the planet are now thought to be an illusion created by the very deep ravines on the planet. *(Hale Observatory Photo)*

The Planet Jupiter. The giant planet of the family, Jupiter, is here seen in greater detail than in any previous photograph.

This view is from a distance of nearly 3-million kilometres — about ten times the distance of the Moon from Earth — yet, it still shows the Great Red Spot and details of Jupiter's cloud formations.

(NASA photo)

rugged terrain and the biggest volcano in the Solar System. Pitted, broken, as heavily cratered as our Moon, the surface contains a great rift, with side walls as high as Himalayan mountains, long enough to cross the entire width of North America — and deep enough to make the Grand Canyon seem a side gully. Spacecraft have revealed no sign of the legendary Canals, but these may have been illusions caused by the very deep ravines. Mars' atmosphere is much thinner than was thought; and there are two small moons, Deimos and Phobos.

JUPITER: With a mass greater than all the other planets combined, Jupiter is the family giant — and a puzzle. Five times the distance of Earth from the Sun, at 778-million kilometres, it spins on its axis (tilted to 3 degrees) every 10 hours and takes nearly 12 Earth-years to complete one solar orbit at a speed of 13 kilometres per second. The giant planet has a diameter of 140,000 kilometres and a mass 300 times that of Earth. Its density is much less, however, and it fills a volume 1,312 times that of man's world. Jupiter has an atmosphere of hydrogen many hundreds of kilometres deep and its main structure is frozen gas with — it is suspected — a small core of rock and metal. This combination is linked with far ranging floods of electrons in an enormous magnetic field. Raging storms occur in the highly turbulent atmosphere. The two Pioneer Spacecraft have transmitted information which suggests the famous Great Red Spot may be dust, from an eruption, which has hovered in a tornado still blowing after hundreds of years. The major planet has 12 moons — one of them, Gannymede, as large as Mercury, and two others bigger than Earth's moon. Some of these satellites are under close study for what they can reveal about the puzzling Jovian environment.

SATURN: The second largest planet, Saturn, bowls along on an orbit 1,426-million kilometres from the Sun in an annual circle which lasts 29½ years. Its orbital speed — 9.64 kilometres per second — is a little more than 3½ kilometres a second slower than mighty Jupiter. Not that Saturn is small; it has a diameter of 115,000 kilometres and is also mainly a huge ball of frozen gas with a density below that of milk, but slightly above water. Saturn was the outermost planet known to early astronomers; and its rings are a dominant feature. Believed to be frozen dust and ice crystals caught in the planetary gravity, they float in orbit in three rings ranging out to 77,000 kilometres from the surface. Saturn is very, very cold; its surface temperature is believed to plunge to minus 180 degrees Centigrade, 50 degrees colder than icy Jupiter. There are ten known moons, the largest being Titan. Saturn's axis is tilted from its orbital plane to about 27 degrees, so that it has winters and summers, each lasting about 15 years.

Saturn is mainly a huge ball of frozen gas. Its rings are believed to be frozen dust and ice crystals caught in the planetary gravity.

(Photo: Hale Observatories)

URANUS: Unknown to man until its discovery in 1781 (by Herschel), Uranus is twice the distance of Saturn from the Sun — 2,869-million kilometres out and lumbering along an orbit which takes 84 years to complete. Nowhere near as big as the two giants, Uranus has a diameter of 47,400 kilometres, and its orbital speed is much slower — down to a little less than 7 kilometres a second. Uranus spins on its axis about every ten hours, but, that means nothing in terms of day and night, since the frozen sphere virtually rolls round the Sun on its side — its axis lying at 98 degrees to the plane of orbital rotation. This means, also, Uranus has a summer and winter lasting 42 years each, except that those terms are meaningless on a planet where temperatures remain down about minus 190 degrees Centigrade. With much less frozen hydrogen than Jupiter or Saturn, Uranus appears to be composed mainly of light elements, mainly helium, nitrogen, carbon and oxygen. Uranus has five known moons.

NEPTUNE: This planet lies almost twice as far out from the Sun as does Uranus. It follows a path some 4,500-million kilometres from the solar centre, a far-flung orbit which takes the planet 165 Earth years to complete. Seen from Earth as a pale green orb, Neptune's existence was unknown until its effect on the orbit of Uranus was detected in the mid-1800's. While the tilt of axis against the orbital plane is 29 degrees, much less than Uranus, Neptune's rotation conforms to almost normal. The orbital speed is 5.4 kilometres a second and a day on Neptune lasts 15 hours 40 minutes. Little is known about the planet's composition except that its chemistry is probably similar to Uranus. Its surface temperature cannot be much above minus 250 degrees C., and it has two moons. One, Triton, is as big as our Moon and, probably, is formed mainly of ice.

PLUTO: This small, lonely planet lies out some 6,000-million kilometres from the Sun and revolves round the edge of the Solar System on a remarkable journey on which one orbit lasts 248 Earth years. Its path is elliptical — not nearly circular like the rest of the planet family — and is inclined 17 degrees against the general plane of the paths of all the other planets. Pluto moves through icy dimness at the lowest speed of all — less than five kilometres a second. Since it was discovered in March, 1930, we have seen it travel less than a fifth of a single orbit, and its erratic path may yet provide more surprises. As with Neptune, Pluto was also predicted, and then discovered, because of its effect on the inner planets. Pluto's period and angle of axis are not known with any certainty, neither is its chemical make-up. Its density is reckoned, however, to be close to that of Earth, while its size may be nearer to Mercury.

Close up of a Giant Planet. The nearest photograph of Jupiter was taken by Pioneer 11 — from a distance of 42,000 kilometres. The view is of polar regions of the massive planet on its northwest shoulder which are not visible from Earth.

(NASA photo)

Viking to visit Mars — Centuries of speculation on the red planet, Mars, may soon be at an end. American space scientists hope the planned Viking landing (pictured in this artist's conception) will show conclusive details of any possibility of life on Earth's companion planet as well as give data on the atmosphere and the Martian terrain and structure. Viking will achieve the first landing of a package of instruments on the surface of Mars in 1976.

(NASA photo)

NEIGHBOURS OF THE SUN

Coming to terms with the scale of distances and life-times of the stars in our sky means adjusting our minds to new dimensions. One fact alone makes this point vital to understanding of space. It is that in this age of technology the Sun's nearest star neighbour is still unreachable.

Though it swings each night into the southern sky, bright, clear, easily identified — and on the galactic scale, this star is right in our cosmic backyard — we cannot yet visit it. Only by intellect and logical observation can we tap its secrets, study its nature and habits. Only from afar can we probe a world where conditions are quite alien to those of the star round which we live.

Not until a future age, when nuclear-powered automated spaceships, on that first great inter-stellar odyssey, are approaching our sun's closest star neighbour, shall we obtain exact details of that different world. It will be an expedition across 40-million-million kilometres of dark space, through a vast void, empty, but for the rare, thin gas and dust of the so-called interstellar medium.

Yet, the presence of this material and the abundance of the elements it contains hold vital clues to the true nature and the future of the whole universe. Such a spaceship will pass through a medium of material so thin it cannot be seen with the naked eye. But, since it fills enormous space it amounts to colossal potential for future star-building.

The gas and dust can gather into clouds — as it has done in the great banks along the plane of the Milky Way — but, all over, its ratios generally average out, with 10,000 atoms of hydrogen to 1,200 atoms of helium, four each of oxygen and nitrogen, one of sulphur, chlorine, neon, and traces of carbon, and metals ranging from iron to gold.

From the work of radio-astronomers we know these materials are there because when they absorb energy from starshine they radiate in distinctive wave-lengths which are recorded on Earth, like chemical "finger-prints", recognisable as coming from specific elements. The radio detectors of modern radio-astronomy show how important this interstellar material is in terms of the creation of new suns.

Our tiny planet makes a single turn round the Sun — and we call that a terrestial year. In that time our orbiting planet completes a journey of 950-million kilometres. In that same period of one year, light from our Sun travels 9.5-million-million kilometres across space — and that distance is less than a quarter of the way to the closest star.

We call that star, Alpha Centauri. Light from our Sun takes four years and four months to reach it; and light travels at almost 300,000 kilometres a second. The true distance — in direct line — is 40-million-million kilometres, but, since that is so cumbersome a sum, we use light years as a measure. Thus, Alpha Centauri is 4.3 light years distant.

This star has several distinctions — other than being near-by in terms of galactic distances. It is a signpost in the Southern Sky, and the third brightest star in all the heavens. It gets its name, Alpha Centauri, as the most conspicuous star in the constellation of Centaur, and with Beta Centauri, second brightest in the constellation, has guided the eyes of millions of human beings to the best-known of constellations below the equator — the Southern Cross.

This first-ever colour photograph of the entire Orion constellation shows the heavenly "Hunter" with the "fiery scabbard" swinging from the three-star belt, in which young stars are now being born. The left foot star is Rigel — some 800 light years distant and 23,000 times brighter than our Sun. The right shoulder star is the giant red sun, Betelgeuse.

Time and the stars

The star-pattern of the night sky can have changed little since the dawn of human time. Some of the oldest stars have vanished while new ones have been born; some have exploded and left their remnants strewn across space. In the well-known constellations some stars have drawn marginally closer — some have drifted apart — and the Moon itself has drifted an inch or two further away. But, overall, the sky looks much the same to our eyes as it did to those of Java Man — and *his* archaic ancestors.

This doesn't mean the universe is unchanging, or that it is everlasting. What it says is man's time on this planet has been a brief interlude on the timescale of the stars. It illustrates, clearly, how our sense of time is out of joint with the great Milky Way galaxy which created our world — and us along with it! Like a hand on a clock face, the Solar System turns around the galaxy — just as Earth does round the Sun — and since Sun and its planets were created, some 5,000-million years ago, they have made only 20 revolutions round the Milky Way. It is on that almost inconceivable time-scale that the whole of human history seems but a tick of the galactic clock.

Known commonly as The Pointers, Alpha and Beta Centauri line up to indicate the topmost of the five stars forming the Cross. In that role they seem to ride side-by-side, like sister stars, or twins, linked in the eternal task of leading human eyes to the glory of the Cross. But, they are an illusion.

We know that all light moves at a constant speed; and, since Alpha Centauri is the Sun's closest neighbour, we know its star-shine reaches us ahead of light from any object out in the universe. We know the light from this nearest star to the Sun is 4.3 years old when it reaches our eyes, because it takes that time to cross the void. Part of the illusion is in the other Pointer. The light we see at the same time from Beta Centauri is aged by comparison, for, this seemingly sister star is a distant stranger. Beta Centauri is far-off, a remote 300 light years from Earth. Yet, from its position nearly 296 light years beyond Alpha Centauri, the innermost Pointer shines almost as brightly as the Sun's closest neighbour. It does that because it burns with a brilliance 5,000 times greater than the Sun, against Alpha Centauri's 1½ times.

And Alpha Centauri itself contains illusion. Its light, which we see as a single point at casual observation, resolves in good telescopes into two separate glowing stars, two yellow objects shining out together. These two stars circle about each other once every 80 years — and that fact eradicates any idle thoughts of finding planets there which might harbour some forms of life. Moreover, in recent times, a third smaller star has been found to be part of this system. It has been given the name *Proxima* Centauri because its wide orbit brings it for part of its time closer to Earth than the two shining stars it orbits.

All three, as part of a system, were probably born from the same cloud of interstellar material; and all three are changing their position in our sky. Some 2,000 years ago Alpha Centauri was some 12 degrees further to the north and could be seen by Egyptian astronomers in Alexandria where they gave the name of the man-horse creature — Centaur — to the whole constellation. Now, along with Southern Cross and other main southern features, it cannot be seen overhead above 35 degrees south, which is the latitude of the largest southern cities, of Sydney, Auckland, Perth, Montevideo, Buenos Aires, Santiago, Valparaiso.

Future generations in all those places will see the gap between the two Pointers gradually close as Alpha Centauri swings down toward the South Pole.

At present Alpha and Beta Centauri (and they surely deserve special names of their own) line up to point to the topmost star in Southern Cross, to the star known in astronomy as Gamma Crucis. By 4,200 AD that line of direction will point at the midriff of the Cross; and the space between Alpha and Beta will be half of what it is now. In a further 2,000 years the movement of Alpha will have closed the distance between the Pointers. Seen from our planet as a single point of light, they will form the most brilliant and conspicuous object in the whole sky, apart from the Sun and Moon. Two stars of the first magnitude conjoining to shine together on Earth is a very rare event and will not happen again for enormous time. For people on Earth in 6,200 AD the Pointers will outshine Southern Cross, but Alpha Centauri will still be our nearest star neighbour.

In the complete heavens — north and south — about 8,000 stars can be detected with the unaided eye. Only two of these outshine Alpha Centauri in our sky, and both of these giants frequent the south. Shining brighter than both Pointers in our eyes is Canopus, the white-yellow diadem of the southern constellation of Carina, part of the original Argo group, and which rides against the background glow of the Milky Way, below and to the right of Southern Cross.

Canopus holds a place in the history of human exploration of the Solar System. In the lunar Apollo missions which opened with the landing of Armstrong and Aldrin on the Moon's Sea of Tranquility, in July, 1969, Canopus was a critical navigating star. Canopus burns with fierce luminosity — 1,500 times brighter than the Sun. But, distance dims its brightness down to second place among the stars seen from Earth. Pride of place as the most brilliant star is held by Sirius, in Canis Major (Large Dog). This sparkling orb in our sky is a local inhabitant and is not in the same league of actual brightness as giants like Canopus. Sirius lies about twice the distance of Alpha Centauri at 8.7 light years, but holds the distinction of man's brightest star by shining with a power 23 times greater than the Sun.

We can easily find Sirius by finding Canopus. A line through the top star to the bottom star of the Southern Cross carried on for a little more than three such distances locates the eye on the South Celestial Pole. Right from there — at right angles — brings the eye to Canopus. Sirius will be found a similar distance further on in the same direction. Sirius, also, has a place in astronomical history in that it has a smaller and dimmer companion star which was predicted to exist long before it was finally located.

Like Alpha Centauri, Sirius is a so-called "binary" system; and as we look through the universe we find many such pairs, or triplets, of stars in association, and groups leading up to the galactic and globular clusters, powerfully luminious and holding as many as a million stars. Three supreme examples of these are in the Southern Universe.

Several such systems occur in the region round the Solar System — the space-sphere

THE NEAREST STARS

which the great American astronomer Walter Baade once called "our local swimming-hole" among the stars. Suppose we cast our locality out as far as 16 light years — that is a distance of 150-million-million kilometres in all directions. What do we find? A total of about 50 stars which, like most star populations, have variety in size, colour, surface temperatures (therefore brightness) and in age. There are few giant stars, that is stars which seem big compared to our Sun. There is Procyon in Canis Minor (Little Dog), at 11 light years distance with more than seven times the solar brightness; there is Sirius itself, and Altair (in the Northern group of Aquila) which, at just beyond 11 light years, is eight times brighter than the Sun.

But, there are no supergiants in our locality, no really big stars of the type which give us indications of the approximate age of local development from primal gas and dust materials, like the great star, Rigel. Here is the scintillating left foot of Orion, the hunter, who strides from north to south each year with his fiery star-birth nebula swinging like a scabbard from his three-star belt. Rigel competes in brilliance with local stars, yet it is at a distance of 800 light years. Because its surface is so awesomely hot, it shines into space 23,000 times brighter than the Sun.

This mighty star could dominate the whole setting of our local "swimming-hole" — but it cannot defeat time. Rigel — against that galactic clock of the home galaxy, the Milky Way — is a fly-by-night. Profligate with hydrogen fuel, it is a vast star candle — burning at both ends . . . very, very fast.

Rigel is a single star, a spendthrift, squandering its fuel resources. Its vivid white light pours into space from titanic conversion of nuclear fuel in its colossal atomic furnace — at a rate in excess of a thousand times faster than that of normal stars. Rigel's ending is written in its brilliance, its colour, its very size. Astronomers write it off as a giant that will fall before completing a quarter of a turn round the galactic clock. Some allow it another 300,000 years to shine, before its atomic pumps finally close down. A short time against the galactic year — for one revolution of the Milky Way takes 250 million years.

How will such a mighty star die out?

Present concepts give it a chance of three fates. It can contract, slowly, endlessly, into a sink of intense density and dark gravitation — dark because not even light could escape — and become a "black hole", a type of object thought to exist in the universe, but, naturally, not yet found.

It could swell up like our Sun will do — into an enormous red giant — and then slip away to a very solid white dwarf — and final oblivion. Lastly, it could provide one of those brilliant cascades of stellar firecrackers which astronomers call — supernova; which means . . . an exploded star! When, and if, that happens Rigel will be so luminous it will shine on Earth in the daytime.

Pyrotechnics on that scale are unlikely in our "local swimming-hole". The 16-light year space-sphere holding our 50 associated suns contains no supergiants like Rigel.

The company of stars about our Sun are, in the main, slow burners. The majority, averaging in size just below or above our Sun, fall into the class of objects astronomers log as "main sequence" stars.

Most of the Sun's companions are logged among those stars which conform to steady existence. They include the few large stars already mentioned and while these are the fast-livers of the stellar family, the rest are mainly old, tired, red dwarfs, depleted and fading, and orange or yellow stars burning their fuel so slowly they will go on shining billions of years after Rigel, Canopus, and Sirius have all burned themselves away. One such slow star in the Lupus (Wolf) group — which rides the southern sky between Sagittarius and Centaurus — is only 7.7 light years from us yet brings no more than a faint blush to its sky; it is some 60,000 times dimmer than the Sun, but it will in human terms burn for ever. This Wolf star is so small it might be a companion star which slipped away from a former associate. That would not be unexpected among our local "fifty", for, there are a half-dozen small stars which give indications of having companions that might be other small suns or — or large planets!

There is Barnard's Star a northern object for which planetary claims have been made. It is suspected of having at least two planets, one in that critical belt of life-probability which Earth occupies round the Sun — but, these claims are still treated with doubt and suspicion, and occasion some controversy. We do not yet *know* of any star with a family of planets, although world-rank astronomers assert that of the estimated 100,000-million stars in the galaxy one in every hundred could support a planetary system. (This poses the possibility of there being 1,000-million planet-suns in the Milky Way system).

In our "local swimming-hole" we have not yet found a single planet other than our Sun's own family. Could we, then, perhaps in compensation, find a sister star to our Sun?

There is very little chance. We shall never know whether our creator star was born in isolation or in companionship. Time and distance defeat such detective work in space.

Our nearest star is two. Alpha Centauri, the outer Pointer, nearest star in space to our Sun, shows up as a binary system in this telescopic photograph.

(Photo: V. Ford — Mount Stromlo.)

Suppose our creator star *was* born in companionship? And early in our local creation the sister star started to edge away . . . say, at only a single kilometre a second (one-fifth the orbital speed of the slowcoach planet, Pluto). Where would it be now? Professor Bart J. Bok has calculations which show such a star would now be near the edge of our 16-light year swimming-hole; and moving still further into space. We cannot even be sure that any of the local stars had common origins. They may well have drifted like travellers and come together at this point of time — when man is on Earth to observe. That could well have happened to any of our Sun's neighbours; for, this whole cavalcade of nightly glory is moving, wheeling, turning in space — about the hub of a galaxy, about a region which we have never seen, and never will see.

The Southern Cross. Best-loved and best-known constellation of the heavens in the south, with the two Pointers (left) clearly indicating the five stars of the Southern Cross. The outer Pointer, Alpha Centauri, is the nearest visible star to the Sun. This picture shows the rich star-fields of the Milky Way disk behind the Cross.

The Southern Cross

The stars that form the Southern Cross lay greater claim to human hearts and minds that any in the teeming Milky Way.

In all the starfire of the immense galaxy, these are five points of light which form a small constellation; yet they comprise a stellar diadem adorning the South Celestial Pole and mark the point on which the Southern Universe revolves.

Together, these five stars, visible to the unaided eye, are the best-known, best-loved in the great southern sweep of creation; and each of the five is a jewel in its own right. They sparkle with colour that travels for many, many years to reach our eyes, and, in even moderately-powerful telescopes, they enlarge to a display of precious gems against a velvet setting.

Individually, not one of the stars of the Cross competes in brilliance with great suns, like Sirius, Canopus, Rigel; nonetheless, they hang together in far space in the form of a Cross and, thus, they touch deep roots in human beliefs. Their shape, not their beauty, has won them an imperishable place in song and story, in national pride and love of home. Among the ancient people of Australia, tales were woven around this glittering diadem as it wheeled through their long nights, ever shifting its position above spinning Earth.

In less distant days the Cross would have ridden higher in the sky. Some 2000 years ago it could be seen in the Mediterranean where, low on the horizon, the Greek astronomer Ptolemy, included its stars among his Centaur constellation. The Cross did not gain separate identity until the meaning of Christ spread across the world.

Today, the movement of the Sun and the Earth have carried the view of the Southern Cross much further to the south.

For as long as written records go back, the Cross has served southern seafarers. The unique alignment and position of its topmost star, and the brightest of the five — the star at the foot of the Cross — were noted early as a guide to travelling men.

The Southern Cross fulfills this task. Always, no matter the time of year, or the position of the Cross in the sky — on its side or standing on its head — the line from the top star carried through the foot of the Cross for about four times the same distance these two stars are apart takes the eye to the centre of the southern heavens; and, followed down to the horizon, this gives true south.

The Cross can be seen early in the year, gleaming through the twilight from the south-east, resting on its side. The two Pointers, which are unmistakeable, shine like spotlights toward the Cross. These are still considered to be members of Ptolemy's Centaur constellation. Beta Centauri (see Chapter Three) is closest to the Cross, while the other Pointer, Alpha Centauri — being the closest star to the Sun and only 4.3 light years distant — outshines the five stars of the Cross.

By May, the Southern Cross is seen upright in the evening sky; by August, it will appear sloping down to the south-west. It was these movements that puzzled early men and gave rise to imagined gods roaming the nightly vault. The constantly changing setting was easily grasped once we knew Earth rolled around the Sun *and* turned on its own axis.

Place a lighted table-lamp in the middle of a room and walk round it — see how your view of the room changes. Then make the same movement, but, turn on the soles of your feet as you go, and you will understand how the pattern of the universe alters and how at the same point in every journey it reverts to what it was.

A view of the Southern Cross through the Schmidt Camera-Telescope at Siding Springs. The bottom star of the cross, Alpha Crucis, is in the bottom left hand corner of the picture. At any time of the year, regardless of the position of the Cross in the sky, the line from the top star, Gamma Crucis, carried through to the foot of the Cross and followed down to the horizon, indicates true south

(Photo courtesy UK Science Research Council)

From these same movements we can see how some constellations leave the sky and then re-appear later in the year. Some, more directly in the south, never leave the sky. The Southern Cross rides the southern heavens throughout the year from 34 degrees south. It never sets above the cities of Auckland, Santiago, Montevideo, Capetown, Perth, Adelaide, Melbourne, Sydney.

If it was possible for our naked eyes to look through the blue light the Sun spreads through Earth's upper air belts, in daytime, then we could see the Southern Cross move like a giant hand round the face of a celestial 24-hour clock. For, no matter the month, or the year, the line from the top star down through the foot of the Cross for ever and ever points straight to the south celestial pole.

Familiarity with its changing position, month by month, has allowed seafarers through the ages to use the Cross both as a navigational aid and as a timepiece. Long before white settlement in the southern world, the captains of praus from Macassar, in Celebes, used the Cross as a guide-star, conning their ten-ton wooden vessels — with bamboo-slatted shelters, two steering oars and large oblong fibre sails — south to northern Australia in search of the trepang sea-slug, beloved by Chinese gourmets.

In 1505, Ludovico Barthema, a Portuguese explorer, rode on such a vessel as it dipped through the Java Sea. In the tropical night the Macassan captain explained to Barthema how he used the Cross to find true south.

Three years earlier, in February, 1502, Amerigo Vespucci saw the Cross as he ploughed the South Atlantic, and wrote the comment, ". . . to guide the ship, it is much brighter than the northern star." With other travellers of his day, like Vasco da Gama, sailing round South Africa in 1493, he saw the Cross as a reassurance, a shining symbol of his faith in the southern sky.

From the reports of those travellers who returned safely to Europe, the southern constellation took on a quality of mystery and awe in the minds of generations raised in a more devout climate than today.

In 1517, Andrea Corsali wrote of the "guiding cross" in the south; and, 75 years later, its fame led to its inclusion in Molineaux's noted celestial globe. A century later came official scientific christening. Edmund Halley placed the constellation in his southern sky catalogue, and, in the same year, 1679, the astronomer-priest, Augustine Royer, named "this most interesting southern constellation" as *Crux*, the Latin for Cross. The magical appeal of the Cross soon entered literature; even Swinburne, who did not sight the stars of the Cross in his life, "took its beauty on trust".

The writing fashioned pictures, in the minds of young and old alike, of this shining symbol of faith which very few northerners could ever hope to see. One of these was a young lad in Prussia who grew up to be one of the great explorers — Baron Alexander von Humboldt, after whom the main current of the Pacific Ocean was named. In 1799, an eager young man aboard a sailing ship bound for the Spanish Main, travelling with the blessing of King Charles of Spain, Humboldt watched impatiently each night as the vessel blew into the South Atlantic.

Often the lower levels of the atmosphere were filled with tropical sea-mist — and, then, on the night of July 4, all the boyhood memories, hopes and imaginings came to reality. "I saw the Southern Cross for the first time. It was steeply inclined, and appeared from time to time between the clouds, the centre of which in the flickering sheet lightning shone with silvery radiance. If a traveller be permitted to mention his personal feelings, I may remark that on this night one of the dreams of my earliest youth was fulfilled."

Other travellers to the south in those times looked at the heavens with less soul; the French astronomer, Lacaille, listed a number of south polar constellations and gave them unimaginative names, such as Air-pump, Chisel, Octant or Telescope. Not until the mid-19th century was the true beauty of the Southern Cross revealed to men. Sir John Herschel, from South Africa, turned his reflector telescope on to each star in turn and found a stellar magnificence previously unsuspected.

Hanging beneath the left shoulder star — the brilliant white sun astronomers know as Beta Crucis (second brightest of the constellation) — was the star cluster to become known as Herschel's "Jewel Box". Sir John caught his breath and stared in wonder at this vision. He saw — ". . . a collection of diamonds, rubies, sapphires and emeralds laid on dark velvet, like the inside of a jewel box."

Today's telescopes show even greater splendour. The brightest gem, the star at the foot of the Cross, is Alpha Crucis, and this is a setting of two brilliant, blue-white suns, so

Jewel Box in the Cross . . . This is called Kappa Crucis by astronomers, but such is its beauty against the dark of deep distance, that Sir John Herschel who first viewed it from South Africa, named it "The Jewel Box". Hanging below the left shoulder star of the Cross proper he saw it as a "collection of diamonds, rubies, sapphires and emeralds laid on dark velvet . . ."

bright they dim a field of stars beyond. They lie close to 300 light years away.

The right shoulder star of the Cross is Delta Crucis; and it, too, is a doublet of white stars, which are some 200 light years from us. Gamma Crucis, the topmost star, is a brilliant orange-red single sun and is also the nearest of the group — being some 180 light years distant. The fifth star, the dimmest, is set toward the centre of the fretwork of the Cross, and is known as Epsilon Crucis; it is an orange-red star some 200 light years away.

The Southern Cross decorates the flags of Australia, New Zealand, and was on postage stamps in Brazil as long ago as 1889. It gave its name to the aircraft flown by the pioneer aviator, Charles Kingsford Smith . . . and also to a quiet little township, 225 kilometres west of Kalgoorlie in Western Australia which was once a busy gold field.

In 1888, in that far land, Tom Risley and Michael Toomey walked by night and used the Southern Cross to guide them to a line of hills they had heard tell of, from Aborigines. Slightly east of where the line through Gamma Crucis and Alpha Crucis ran to earth, they found the gold they were seeking.

The constellation which gave its name to a town is now more fully explored than ever and its charm is undimished by discovery. Even modest telescopes show the five visible stars to be mounted on a rich field of a dozen groups and clusters crowded with vivid orbs that shine against the remote backdrop of the luminous clouds of the Milky Way.

THE MAGNIFICENT MILKY WAY . . .

Our Sun and its planets are buried among one-hundred-billion stars in the immense Milky Way Galaxy.

No matter where we live on this planet, we always see the white band of glowing gas and stars arching our night sky, because we live *inside* this vast system. And, in all this prodigious creation of giant suns, myriads of stars, streaming banks of hydrogen and helium gas, our Sun is a minor component, our planet Earth a fragment of rock and water. While atoms are the smallest organised form of nature, this system of stars and gas we call the Milky Way Galaxy is one of the largest.

The reaches of the Milky Way are so stupendous even light years become an awkward and inadequate measurement of distance. A light beam from the opposite side of the galaxy races to us at a speed of 18-million kilometres an hour — yet, to reach us this instant, it must have started its journey when Neanderthal Man was still walking this world. It is in this sense, by viewing scenes which shed light so long ago, that astronomers *look back into time*. The Director of the giant Anglo-Australian Telescope, at Siding Spring Mountain, Professor Joe Wampler, says of this point — "A major telescope is really a time machine."

When such a telescope looks across the galaxy, it surveys a scene so vast that light takes 100,000 years to cross from one side to the other!

Much of the matter within the Milky Way Galaxy is compressed into a saucer-shaped disk in which clouds of gas and dust grains, created during billions of years of star-making, mingle and intertwine with an endless array of suns of many colours and wide variety.

Because, from Earth, we see this disk create a circle about the Solar System, astronomers have known for a long time that the Milky Way was flattened. From our side view of the disk they could look along the plane of all that gaseous creation and see stars that were distributed differently to those outside the disk.

Two centuries ago, in the late 1700's, world-rank astronomers such as Sir William Herschel, and the first Astronomer Royal, J. Flamsteed, saw our galaxy-universe as "watch-shaped" . . . but, surprising as it is to our minds today, they had our Sun featured in the centre of this creativity with the star numbers gradually diminishing toward the edges of their stellar timepiece.

They were presented with problems from this concept. The deceptive motions of stars, and of our Sun, made the solution well nigh impossible without more modern visual aids — and much insight.

Consider the conundrum astronomers faced, located in this colossal disk looking out of it in one direction, into its depths in another, seeing only a tiny fraction of the whole of this creation! How could they probe and establish the true size, shape, and form of the stupendous galaxy? Yet, it became possible, eventually, through the keen eyes of a woman astronomer. She opened a pathway to important discovery and an enormous leap forward in human understanding of the whole cosmos.

Henrietta S. Leavitt was an assistant at Harvard Observatory, and, in 1912, she studied the southern sky from South Africa to look at a class of star known as "Cepheids" — one of several types with the puzzling ability to vary their brightness. They were called variable stars. They were to be highly important to world astronomy.

By then, stars had been classified into groups, and systems of matching brightness against known distances were being developed slowly. These were the astronomer's

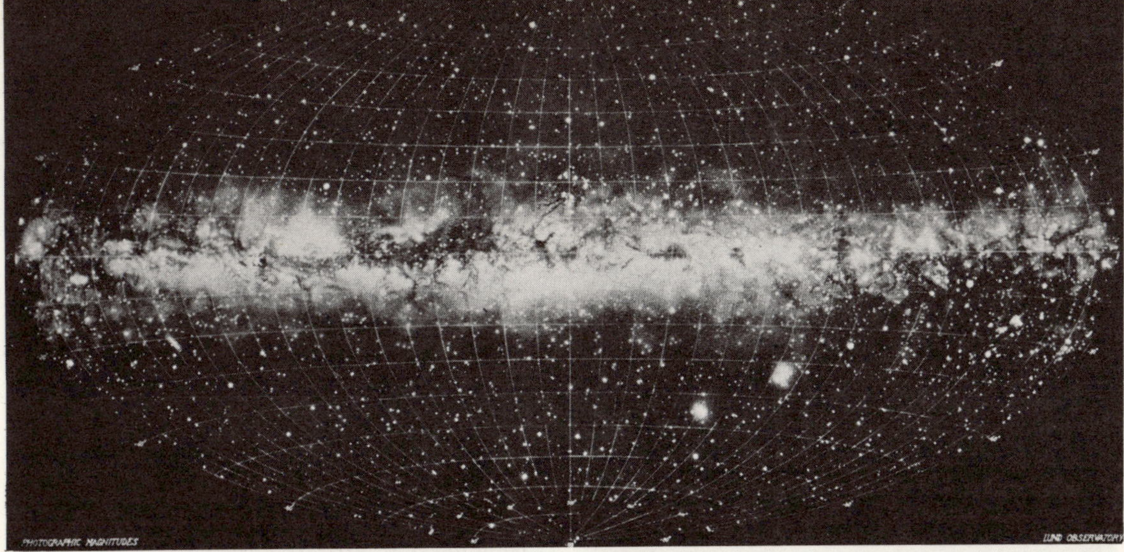

Milky Way Galaxy. If man could leave the tremendous galaxy which is home to the Sun this is the view he would see from outer space of the magnificent Milky Way. It would look like this in his eyes, in the narrow band of visible light to which human eyes have evolved. But, this is not a photograph so much as a map of the distribution-from edge-on — of the material, the stars, the great Nebula and globular clusters of early stars along the plane of this stupendous system. The two white blobs at lower right centre represent the nearby smaller galaxies, the Clouds of Magellan, and the Sun is buried along the gaseous plane two-thirds of the way out right of centre. In the central area of the flattened disk are the star-rich clouds behind Sagittarius which hide the gleaming bulbous heart of the Milky Way Galaxy. All along this plane the brilliant stars illuminate the massed clouds of hydrogen and helium. This map was drawn painstakingly by two Swedish astronomers at Lund Observatory, Titania and Martin Tesküla. Stars are to scale of luminosity; the Sun is so small it is lost in this crowd.

most promising methods of dealing with the difficulty of determining distances between stars of apparent size and brightness, but, obviously, much different because one or other was very remote.

It was just such a case as we have seen with Alpha and Beta Centauri, appearing to be linked but separated by 296 light years. True size and intrinsic brightness were vital objectives.

The "Cepheids" Miss Leavitt studied in the southern sky were found in several features; she scrutinised their behaviour meticulously, recording the times and periods of fluctuation in brilliance. And, poring over her carefully compiled results from the south, a light dawned in her mind. She saw that in the Cepheids the periods of fluctuation matched closely the brightness of the star. The finding was to prove classic. It sent a flurry through world astronomy. Variation periods revealed intrinsic brightness — and the way was open to determine size; and so — true distance.

For the first time man had space "candles" he could use as milestones to measure his great galaxy.

There was no sudden jump into blinding realisation; many years of precise correlation had to take place. There were many arguments, in public and private; but, by the mid-1920's the space "candles" in great globules of clustered stars — lying outside the main plane of gas and dust — had yielded secrets of distance. The American astronomer Harlow Shapley pictured the flat shaped galaxy with a centre far from Earth. He located it behind the constellation of Sagittarius which rides overhead in the Southern Universe for much of the year.

Cepheid "candles" and the globular clusters led to Shapley's galaxy; and, his findings led on to joint agreement that this huge collection of gas and stars rotated around the hub that was located behind Sagittarius. Later, it came to be understood that the motion of rotation was a mechanism preventing further collapse onto the hub of the Milky Way material.

This implied that sometime in the past there had been a collapse to the current formation. When? The present Director of Mount Stromlo, Professor Olin Eggen, Dr. Alan Sandage of Mount Palomar, with an English Associate, Dr. Lyndon Bell, worked this problem to conclusion by the early 1960's. They found the Milky Way had originally been shaped like a ball — a sphere of gas formed of original primal material which had structured the galaxy.

The oldest object in our sky. This teeming family of at least ten million stars is a globular cluster known as 47 Tucanae. Older than any object we know in the Galaxy, bigger and nearer than any of the 200 such clusters in the Milky Way, it is here seen in greater depth than in any previous photograph. This view was taken with the giant AAT telescope on Siding Spring Mountain. Located in the southern constellation of the Toucan it is a mere 14,000 light years from Earth and is judged to be 10,000-million years old.

In this picture there are visible stars which are 100 times fainter than the Sun and millions that are very much brighter. This venerable creation is now being carefully studied as a past chapter of the time when our Milky Way Galaxy was created.

(K. Freeman, AAT, Siding Spring)

While in this shape, in the first few million years, the great star colonies called globular clusters were formed. Since many of these great slow-burning stars still exist (and are of high importance to modern astronomy) they were to help reveal an age for the galaxy.

The Globular Clusters — three of which in the Southern Sky* are closer and more revealing than any in the north — became astronomy's Old Men of the Sky . . . the antiquities of the Milky Way.

Eggen, Sandage, Lyndon Bell found the galaxy had collapsed after about 500-million years and that these early stars were left outside, pursuing erratic orbits about the Hub in some cases, but, together, forming what is now recognised as the "Halo" population of stars, among them the most ancient objects in the known universe.

Discovery of the history of collapse, the shape and size of the Milky Way and its rotation left further problems to be resolved. Looking into the great clouds behind Sagittarius, peering either side of the world into banks of obscuring material gave man a picture of a flat galaxy — but that was the side-on view! What was its true shape? Was it a rounded flat disk, or were there arms spiralling outwards as many astronomers suspected? This could not be proven until the advent of radio-astronomy, which was born into the late 1940's out of war-time radar research?

Interstellar medium, the gas and dust (as we have seen between the Sun and Alpha Centauri) reveal their presence, density, speed, to listening radio-telescopes on Earth. They radiate in known wavelengths, either from their own intrinsic energy or by absorption of energy from other radiating bodies. Their radio "fingerprints" unfolded astonishing details of our galaxy — indeed, the whole universe. The radio studies done on the galaxy, mainly in Australia, and by Dutch astronomers resulted in the compilation of a map of the Milky Way. It is something optical telescopes can never see, because, while radio wavelengths can penetrate dust and gas clouds, the shorter wavelengths of light are blotted out.

The enormous leaps forward in recent times have now given man a picture of his local universe — the Milky Way creation. It is a picture that has strong validity, because (as will be explained later) we can see it repeated across the vast outside cosmos. It is, however, a picture we may never see with our eyes, for a journey outside and above the Milky Way Galaxy would need an expedition lasting thousands of years.

However, suppose one day man rises to that height; what would unfold to those first human eyes to look down on the Milky Way?

They would see a galactic Ferris wheel — 100,000 light years rim to rim — revolving around a gleaming bulbous nucleus of vast

Footnote: * The three southern clusters are known as 47 Tucanae (pictured), and NGC6397 and NGC 6752; they represent man's most detailed picture of the galaxy's early structure now that they have been photographed by the major telescope at Siding Spring Mountain, Australia.

A coat of gas. Among the star-fields of the Vulpecula constellation is the glowing cloud of gas and dust thrown off by a young emerging hot-blue star which heats and illumines the features known as the Dumbell Nebula.

(Photo: Mount Palomar — 200-inch.)

A ring of fire. Fierce nuclear reactions in protostars forming inside this cloud make the Rosette Nebula quite distinctive; it lies in the southern constellation of Monocerus, between Orion and Canis Minor (Little Dog) and is part of the Milky Way gas band which is highest in the southern sky in January.

(Photo: Hale Observatories — 40-inch.)

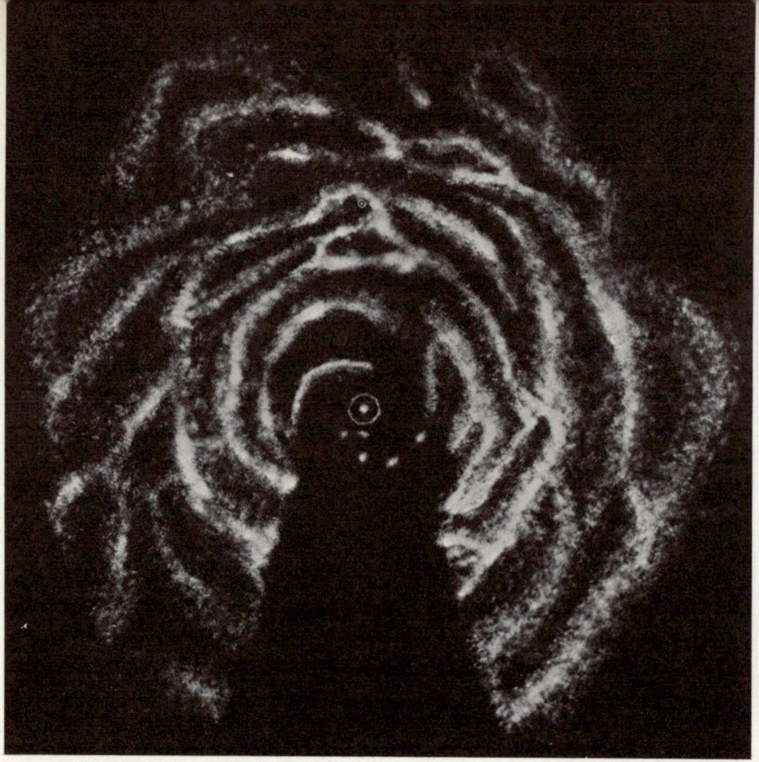

Radio Map of Milky Way Galaxy. Spiral arm formation of the galaxy is revealed in this map drawn by a Dutch astronomer from radio signals detected in Australia and Holland from the hydrogen 21-centimetre wavelength emission from galactic material. The precise heart of the galaxy is located by the central circle and the position of the Solar System is two-thirds out towards the bottom of the picture near what is thought to be a spur of a main spiral arm which has been called the Orion. Other spiral formations were called the Perseus and Sagittarius Arms, but direct optical observations of these features depend on "holes" in the interstellar medium, the dust and the grains and the gas, which lie in density waves that may be formed from magnetic funnels of force.

matter, all incandescent and brilliant white-yellow in the light of millions of huge closely-packed suns. Outside this glowing heart the whole scene is alive, moving, shining with liberated energy. The arms of the great Ferris wheel are bent, curved backward to the rotating motion — trailing like fiery plumes from a gargantuan spinning firework. Spiral arms stretch out for 40,000 light years and are sprinkled with glittering stars, old clusters and new, young, hot, blue-white giants. They are lights on creation's grandest carousel. Man rides Earth in one of these arms.

Similarity to the Ferris wheel ends there; the galaxy is a creation too fluid, too vastly varied, too beautiful for true comparison with a rigid structure. The Milky Way is contoured with no cleanly defined rim. The spiral arms are tenuous and uneven. Movement and changing speed runs through the system. Stars closer to the heart of the galaxy speed faster than the Sun and its companions which, together, race on fairly regular orbits at 250 kilometres a second.

It is hard to grasp, reading these words, that our world — which spins on its axis at 1,800 kilometres an hour, and rolls round our star at 108,000 kilometres an hour — is churning through a galactic rotation with the Sun at a further 900,000 kilometres an hour!

Yet, despite this fantastic speed our Solar System needs 250-million years to complete one orbit of the Milky Way, to go once round the galaxy: 250-million earth years to one galactic year. On that scale our Sun, in its 5,000-million years of existence, has made the journey only 20 times. The galaxy — if we judge its age by those venerable clusters out in the old "Halo" of ancient stars — has rotated merely 40 or 50 times since the Beginning.

There is still much to learn: of the sources of power, of the mechanisms controlling the spiral arms, of the history of the great Hub itself, and of the birth, life, and death of the populations of stars and how they contribute to the elements of biological life.

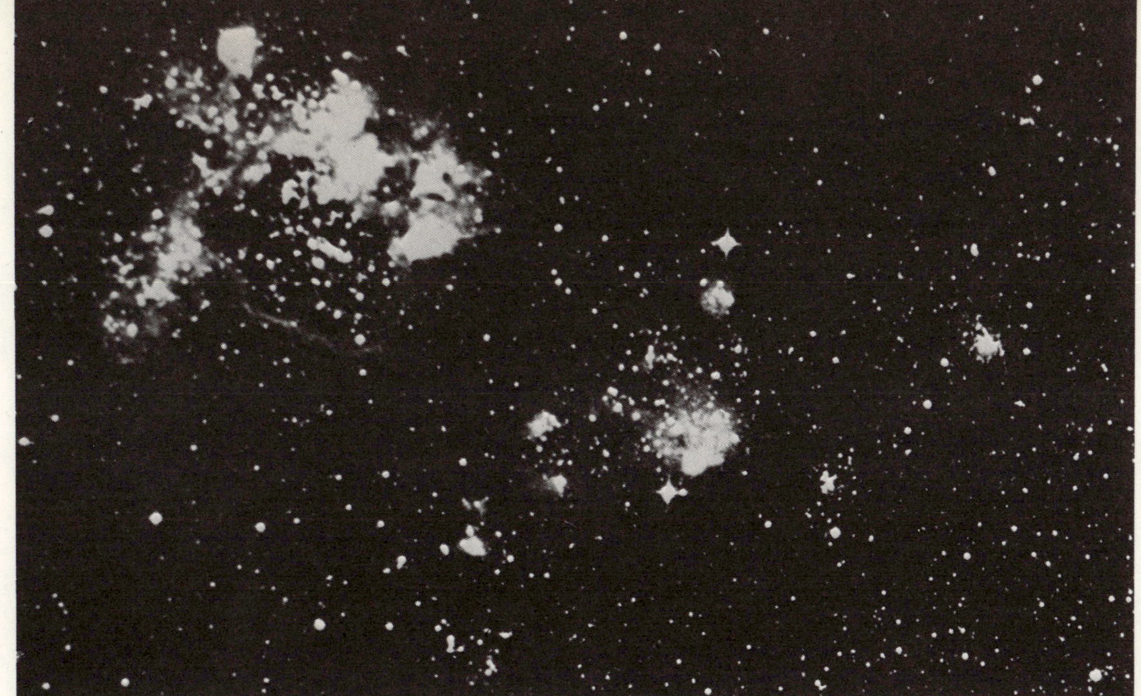

Candles of Space in the Magellanic Clouds. A field of variable Cepheid stars in the Large Cloud of Magellan — such as Henrietta Leavitt noticed from South Africa in 1912. These stars — with their brightness linked with the period of fluctuating luminosity and their true magnitude — were the first standard "candles" astronomers could use with considerable accuracy. From them came the truth that the Clouds of Magellan were galaxies external to the Milky Way — and, eventually, the distances to other far-off galaxies.

(Photo: Prof. Gascoigne — 74-inch Mount Stromlo.)

"Like a swarm of celestial bees . . ." Uncounted thousands of suns live close-packed together in this globular cluster — NGC 6397 — one of the three closest such clusters to Earth. To understand how such collections form, and to know their true structure powerful telescopes are needed to probe down into the globular clusters to examine stars a thousand times less bright than our own Sun.

A STAR CHART OF THE SOUTHERN UNIVERSE

CRATER
SEXTANS
CANCER
HYDRA
Pollux
Castor
ANTLIA
US
CANIS
MINOR
Procyon
PYXIS
VELA
GEMINI
RUX
MONOCEROS
USCA
CARINA
PUPPIS
CANIS
MAJOR
Sirius
CHAMELEON
VOLANS
CARINA
Canopus
South
Celestial
Pole
PICTOR
MENSA
COLUMBA
LEPUS
Betelgeuse
DORADO
ORION
HYDRUS
RETICULUM
CAELUM
Rigel
HOROLOGIUM
Achernar
ERIDANUS
PHOENIX
Aldebaran
FORNAX
ERIDANUS
TAURUS
PTOR
PLEIADES
CETUS
PERSEUS

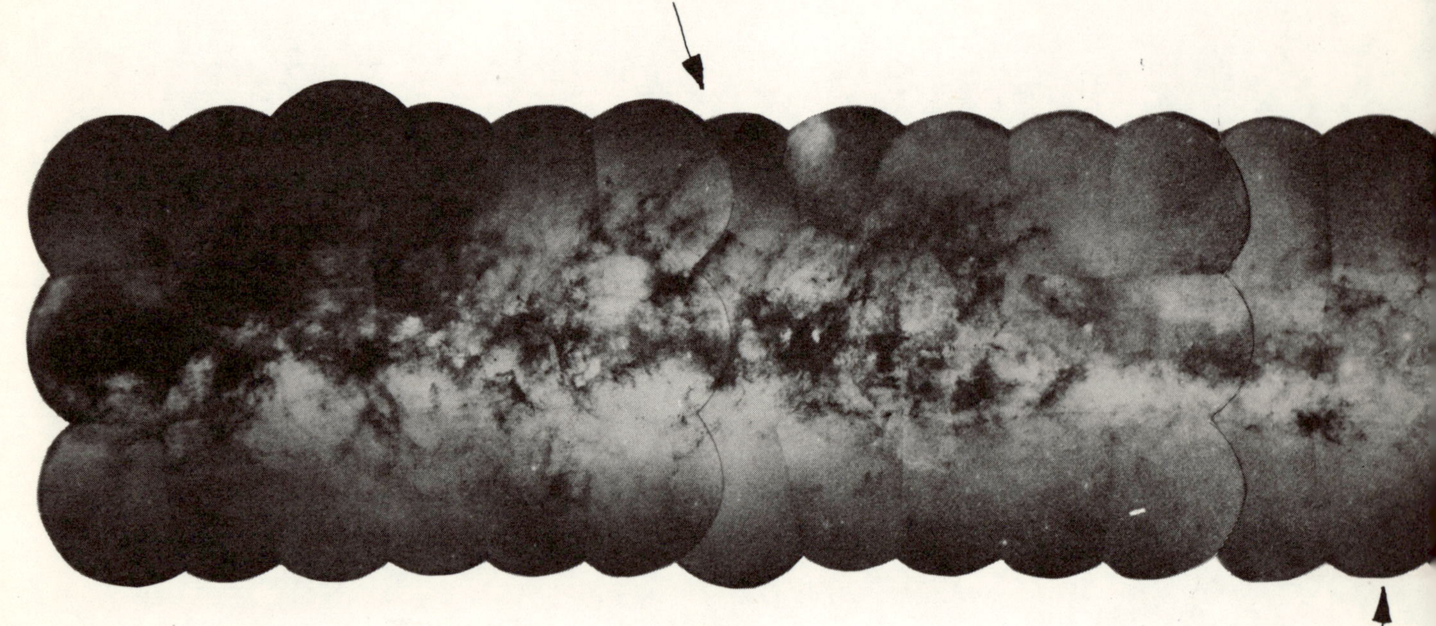

The Southern Milky Way. This remarkable composite photo-picture of the Southern Milky Way was compiled from dozens of telescopic photographs taken from Mount Stromlo and is the only actual complete pictorial scan of the band of the Southern Milky Way Galaxy. At left the dark clouds roll into the Sagittarius region-heartland of the Galaxy; right of centre is the dark smudge of the Coalsack below and to the left of the Southern Cross, with A. Centauri the bright centre star again to the left. In the bottom right corner the very bright star is Sirius in Canis Major.

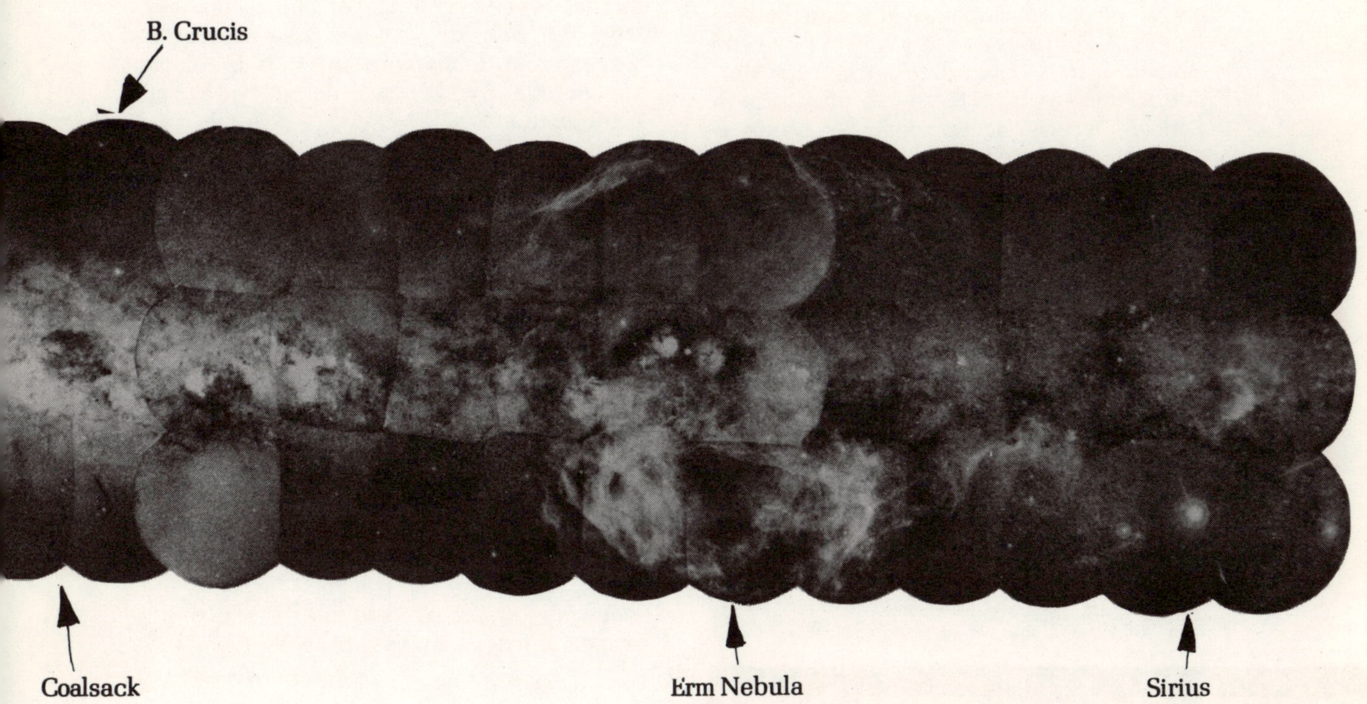
B. Crucis
Coalsack
Erm Nebula
Sirius

Light – the great revealer

All light reaching Earth is packed with valuable information. It carries many clues on the composition and the condition of the material from which it came. It often reveals the presence and nature of the interstellar medium through which it has passed. Light is the astronomers chief yardstick. That is why some of them call it the "great revealer."

All light begins by disturbance in the structures of atoms in stars, or star-like bodies. When such light fills our daytime sky we call it sunshine. It is light from our parent star. When it bounces from the dead, grey face of the moon, we call it moonlight — but it is actually the reflected light of the Sun, as is the light from all the wandering planets — which are often mistaken for stars, though they do not twinkle in the same way. Light from stars we call starlight, and, like sunshine, it is born in the reactions in the stellar nuclear furnaces in which simple hydrogen is broken down and recombined into more complex atoms. The liberated, unwanted material left-over in this process converts to radiant energy which fills surrounding space.

The energy released in these reactions covers a whole range of wavelengths which compose what is called — the Electromagnetic Spectrum. This spectrum starts at one end with very short wavelengths of radiant energy, with gamma rays, x-rays, ultraviolet rays — of intensities and energies that make them harmful to organic life. As the radiant waves grow longer across this spectrum we come to visible light and infra-red waves, and then move on into the longer wavelengths which we detect with radio equipment.

The middle of this spectrum contains the band we know as "visible" light, that is, radiation with the kind of wavelengths which human eyes have evolved to detect. Visible light is the narrow section of the electromagnetic spectrum to which the structure of Earth's atmosphere allows entrance. It enters, astronomers say, by one of two "windows" in our planet's air belt. The other

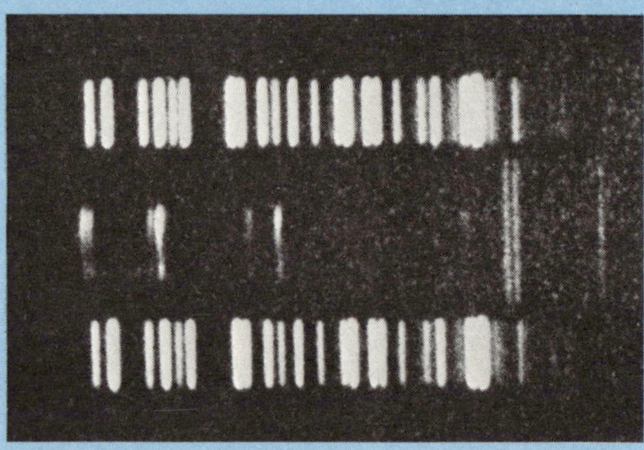

This is the emission spectrum of N49 — remnant of an exploded star in the large Magellanic cloud, 160,000 light years distant. Lines which run right across the centre strip are Mercury lines from Canberra street lights. The narrower (shorter) emission lines show four of the elements in the remnant — oxygen, nitrogen, hydrogen, sulphur.

(Mt. Stromlo — 74-inch.)

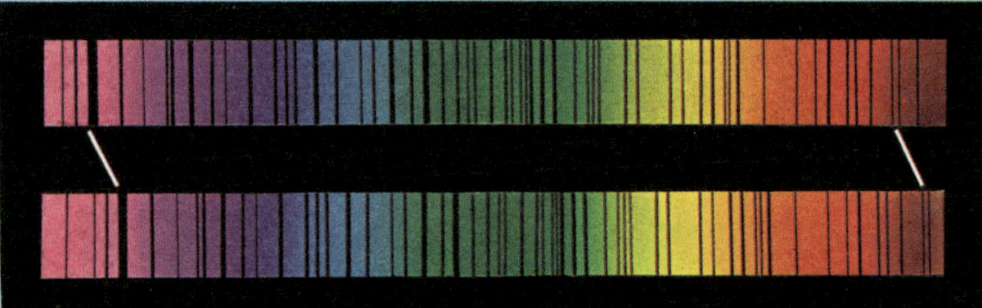

When an object is receding the absorption lines in its spectrum appear displaced towards the red end of the spectrum. This is known as red shift, and by measuring the degree of displacement astronomers can calculate the rate of recession.

When a white light is passed through a prism it can be seen split into its component bands of colours ranging from red to violet. The variations of colour represents the range of visible wavelengths.

"window" is that which admits the wavelengths of radio emission. Human evolution did not cater for that type of wavelength. Man had to discover its existence and invent methods of detection.

Light, however, gives earthbound astronomy its most useful tool. Though it fills only a narrow band of the whole energy spectrum it, too, has its own range of wavelengths. These break white light into separate colours, such as we see with rainbows; and each colour can be quite revealing. Isaac Newton's famous experiments on sunlight, breaking up its colour spectrum with a glass prism, were followed by further discovery; that the shorter and more energetic wavelengths of light created the violet and deep blue end of the light spectrum, and that wavelengths of light grew steadily longer as the colour changed up through green, yellow, orange to red. Wavelengths of red, the longest of them all, were the least energetic — and so proved to be a wonderful yardstick of distance in space.

Deeper understanding of the nature of light came with the invention of the spectroscope and photography. The spectroscope separates light's colour bands and the camera catches the image for long and careful study. Out of this work has come a further feature — the black lines of the spectrum of stars.

Strange as these seemed when they were first detected, they are one of the revealing features of light. They are caused by the behaviour of atomic particles in the atoms which emit the light — or absorb it, as the case may be. Atoms of every known element send a different pattern of black lines. It means that stars transmit their chemical "fingerprints" across space.

Light as a yardstick

The known constant speed of light provides the most suitable measure for astronomical distances.

To speak of the Clouds of Magellan as being some 2.4-million-million-million kilometres away, as the Sun as some 150-million kilometres distant — or even the Moon as at 360,000 kilometres — does not register *real* distance on our minds.

Thus the distance to the Moon is better understood as 1.25 seconds in time of light travel; the Sun is eight light minutes distant — and the Clouds of Magellan some 160,000 light years.

The speed of light over a single year represents 9.5-million-million kilometres, and a single parsec (3.2 light years) a span of some 30-million-million kilometres.

To overcome obvious difficulty astronomers use mathematical shorthand for such long numbers; in round figures a light year would be expressed as 9.5×10^{12}km. The distance to the great Andromeda spiral galaxy would be shown 2×10^{6} light years (2-million); the assessed age of the Milky Way Galaxy might be shown as 10^{10} years — some 10,000-million years. Parsecs can also be used in units of a thousand; thus, 3.2-million light years can be simply expressed as 1,000 kpc.

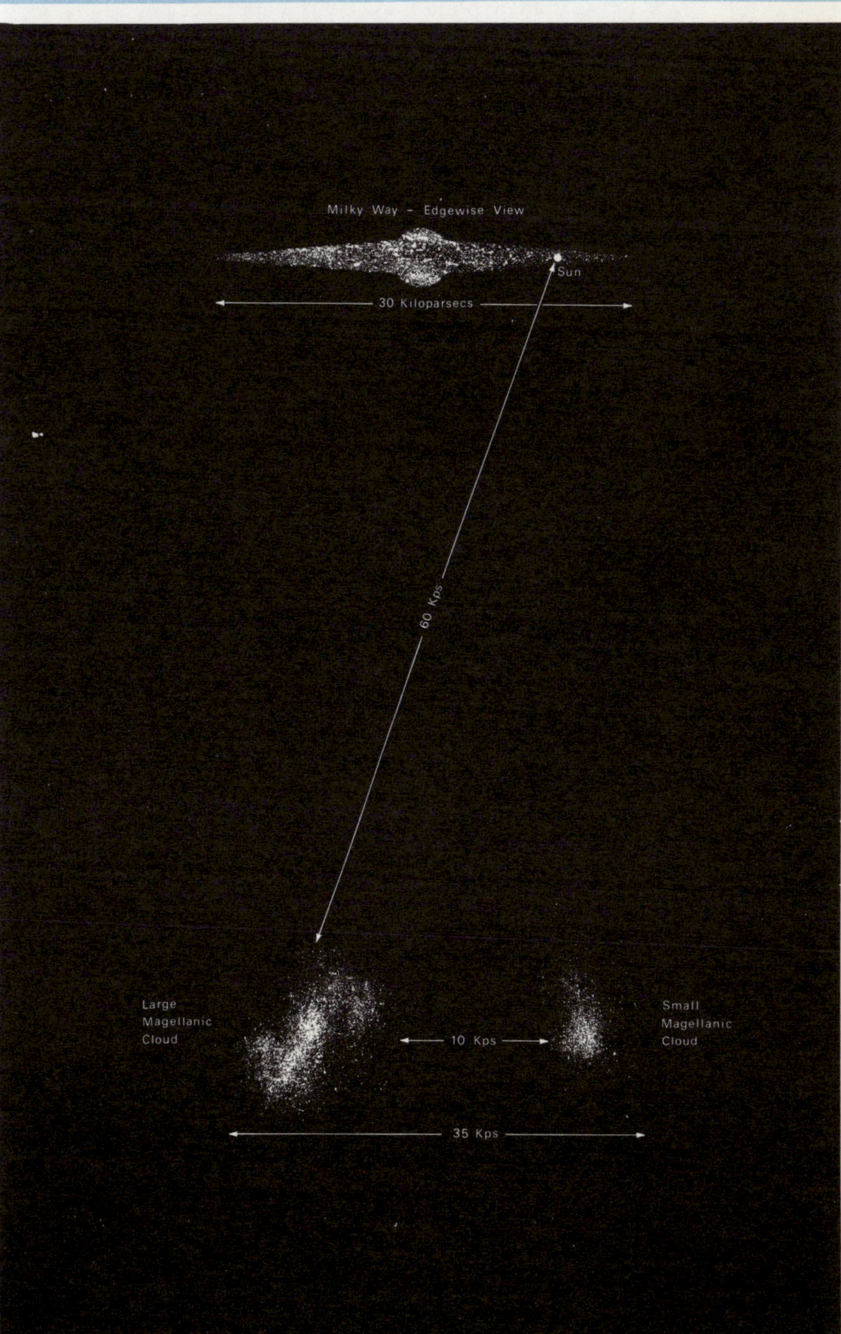

How far to a star?

Look up into the sky at night and, to your eye, each star is a point of light — no more. It may seem bright, it may seem dim, but neither of these facts discloses the distance of that point of light from your eyes. You can make judgments of distance on the lights of a ship at sea, on a car's headlights on the road at night; but, how can you find how far it is to a star?

In the earlier years of astronomical science the trigonomic method of parallax was used; that is, the determination from angles at a known baseline to establish the distance of a far-off point, a method applied by surveyors and architects. This technique, indeed, was used to fix the distance of our nearest star — the Sun!

In 1769 Captain Cook recorded the transit of Venus across the face of the Sun from Tahiti; observers on the other side of the world did the same thing. The angles at the point of contact of the planet with the solar rim, from both observations, were used with the diameter of Earth in this parallax experiment to find the distance to the Sun's face with great accuracy.

Once this distance was known, it became possible to use the parallax technique on nearby stars — employing the known distance of earth's orbit around the Sun as a baseline. Thus, an angle taken on a star when Earth is on one side of the Sun, and another angle on the opposite side (exactly six months later) gave astronomers a baseline of 300-million kilometres for their parallax experiments.

Such measurement of angles, however, are so fine as to be extremely difficult. To fix a point by parallax with angles of only *one second of arc* on either side of the Sun gives a distance into space of 3.2 light years. Distances beyond this demand extreme accuracy and so bring a risk of uncertainty. Beyond the local stars, parallax is of limited use; but it was a first staging post.

The distances staged by parallax were built upon by the patient work of astronomers who carefully sorted and classed the different known types of stars into families. As the knowledge built up of how stars behaved, how some were variable in their brightness periods — such as the already-mentioned

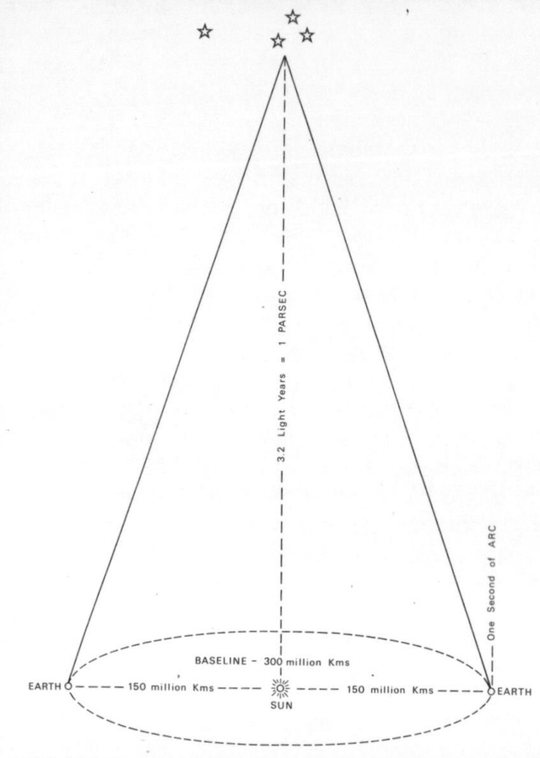

Cepheids and the R. R. Lyrae stars — astronomers realised they were watching stars in different periods of their life-cycles. These accumulated facts also helped with methods of assessing distances.

Later, when the movement of the Sun and its family around the whole great Galaxy became known, it was possible to use parallax on immense baselines, since the Sun travels enormous distances in a single year.

Light itself provided the most valuable clues — not only from the variable stars, but also by the nature of the longer-wave red light in the spectrum. This came out of a finding called the Doppler Shift. It is similar to a car siren approaching you which changes its pitch to lower vibration as it passes you and moves away. So it is with the light from a distant star or galaxy. An object moving rapidly towards us in space shifts its colour spectrum deeper into the blue and violet. One moving away stretches out the less energetic red light waves. The light spectrum moves deeper into the red.

This phenomenon is known as the Red Shift. In the universe the rule seems to be that the further away an object is the faster it recedes, the deeper into the red its colour spectrum is taken. So, the Red Shift has proved to be a most valuable light yardstick in measuring the great universe.

BIRTH, LIFE, DEATH AMONG THE STARS

Down-sky from the Southern Cross, toward the west, blazing Sirius — the Dog Star — shines out from Canis Major. Most brilliant of all the orbs in our heavens, eight light years away and 23 times brighter than the Sun, Sirius is a starting-post on a journey across the staggering creation of the Southern Milky Way.

From a point slightly above this splendid white sun, the glowing track of the Galaxy's flattened disk curves across enormous distance, arching through the white-yellow stars of the constellations of Puppis and Vela, above most of Carina with the unmistakeable Canopus at its foot; past the lovely stars in Southern Cross, with its gleaming Jewel Box, dropping downward by way of Alpha and Beta Centauri — avoiding the tail of the Scorpion — to beyond the stars of Sagittarius, where the piling armadas of clouds hide the centre of the Galaxy.

This centre is the richest stellar promenade of local creation — the wealth of the Southern Sky. When scanned by powerful telescopes the belt of white shining gas behind the bright stars sharpens into a crowded, jostling population of myriads of suns of wonderful variety.

Along this clustered reach of the universe lie some 50-billion suns, one-half of the Galaxy's assessed population. They astonish with their variety. Many great stars — like Betelgeuse in Orion's shoulder — fill a volume at least a million times greater than the Sun and burn their nuclear fuel many thousands of times faster; many orange suns, smaller and larger than our slow-burning star, glow into space. Others are bright yellow, some blue and green, some heated white, like Sirius. All are part of the Milky Way creation, for, the major product of the Galaxy is stars. They are evolution in the sky, a pattern of celestial development in which each shining face tells the astronomer a story of the star's nature, of its future . . . and the future of the Milky Way Galaxy itself.

Brilliant open and closed clusters, born in common clouds of stellar material, spangle the Galaxy's disk. Stars that are alone, stars that circle each other in twos, or threes, globular creations compressed into tight spheres of a million suns; all obey the immutable laws of astro-physics. For all their astonishing variety of mass, colour, magnitude, they all relate the same fact to observers on Earth. The unchangeable fact is that stars cannot last for ever.

Like populations on Earth, they are born, they live out their cycle, and then they die. Some will slip quietly into old age and fade away to the dark cinders which may well feed the gas clouds with grains of star dust. Some will flare out in a desperate straining to hold on to their flame of existence; a few will die violently. But all will vanish — and some will be replaced by other suns . . . so long as there remains the wherewithal for building new stars, so long as the Milky Way Galaxy can call up reserves of hydrogen to light nuclear fires in the young stars of the future.

The lesson is plainly written across the crowded sky. If there is gas and dust — there is star-making going on. For these hazy, glowing areas astronomers use, the term nebulosity. However, these star-birth clouds can also be dark, and dense enough to obscure the fields of background stars. Professor Bart Bok has called these dark sky areas . . . "Holes in Heaven."

Some of this activity may not be so far away. Just east of the foot of Southern Cross is the Coalsack. Heaving its dark shape into the glow of the Milky Way, this is an excellent example of a star-bearing complex. It covers the sky for five degrees square (about 12 light years in diameter) immediately next to the Cross; and while its nebulosity is thin enough in places to see stars beyond (though these are reduced in brightness by

A head of activity. A vast cloud of star-making gas and cool star-dust forms the mighty Horsehead Nebula. Rolling upward from the stretch of nebulosity in the belt of Orion, the cloud is darker because it is cooler, but it is also prominent because, beyond and among, the bright young stars in this great field are supergiant blue-white suns which make the more distant clouds luminous. More than 500 light years away, this area of the sky is closely watched by observers of stellar life-cycles.

(Photo: Mount Palomar — 200-inch.)

The Great Nebula in Orion presents a magnificent spectacle, with great arms of gas streaming from its brilliant heat.

up to 15 times) there are a few darker globules of dense material that really do make "holes in heaven."

The Coalsack has been intensely studied by Dr. Alex Rodgers, of Mount Stromlo, and interest is high because the cloud is relatively so close to Earth — only 500 light years away. It is part of our local spiral arm, and, where telescopes can see through the gas and dust of the Coalsack, there is nothing beyond for 2,500 light years, across to the next spiral arm, the one which follows ours in the rotating Galaxy. Dr. Rodgers estimates these dark cores hold material 300 times as dense as the surrounding gas; that the whole Coalsack cloud probably contains enough material to form 20 new stars similar to the Sun.

It seems the Coalsack is a stellar nursery — a section of the overhead southern sky where baby stars are soon to be born. But, aeons of time are involved in stellar natal processes, and, while modern astronomical equipment can find stars across the firmament in various stages — from glowing nebulosity to young stars shining through their maternal clouds — human science has not been existent long enough to follow the actual birth of a star.

The prospect of new stars being born in our corner of the Milky Way Galaxy is exciting to southern astronomers. The Sun and its planets ride some 33,000 light years out from the Hub of the Milky Way, which is directly behind Sagittarius. And, out in our part of this gaseous disk the material for future suns is getting sparse. We must remember that our middle-aged Sun is a fairly old resident on the galactic scale and that in 5,000-million years the gas has been depleted locally.

Recent work shows the local interstellar medium is down to about one atom in each cubic centimetre of space — that the gas disk where we live is a mere 1,000 light years thick. Such conditions make the adjacent Coalsack (which can be seen clearly on a dark night with good binoculars) valuable to astronomical theories.

Astronomy, of course, still rests heavily on theory. The great concepts about the Universe are supported by a small base of actual fact, the kind of data we can test empirically on Earth. We have not followed a star's birth; neither can we ever travel to witness the actual process of energy production in the Sun's interior. But, science does not need to see an atom to know it is there. Methods and exclusive techniques can vet a

Birth of stars. Young suns sparkle with new brilliance near folds of clouds where star-making still goes on. This is part of the Lagoon Nebula in the Sagittarius region of the Southern Milky Way.

(Photo: Mount Palomar — 200-inch.)

proposition and reject ideas which do not fit known facts. Today, we can build numerical "models" of stars for computer analysis and trace their life-cycles from beginning to end.

Results from this work match the evidence in the sky — on how stars are born, shine their time away, and then blink out.

The profuse variety of star types, the brightness and colour, caused confusion among astronomers for a long time. Why did one star burn brilliant white, another blue, or green, and others red or orange? In the first few decades of this century the answers started to appear. Brightness and colour were linked with size and age. It was this realisation which made it possible to class the stars into related groups and to produce the Hertzsprung-Russell Diagram (already mentioned) which has become a widely-used star chart for astronomers. Another system relevant here is classification of stars into "spectral" types, a system which groups them together by their apparent luminosity.

There are nine main Spectral types of stars, and these are indicated in astronomy by the letters — O B A F G K M N S.

Tutors of astronomy have found that when this is paraphrased to read — "Oh Be A Fine Girl, Kiss Me Now Sweetly," it somehow sticks clearly in the memories of students!

O-type stars are the very hot ones — with surface temperatures about 50,000 degrees Celsius — while B-type stars get down to around 15,000 degrees C, and the coolest, N-stars, have a surface heat of about 2,000 degrees C. Each of these types, however, can be sub-divided into sequential groups. For example, our Sun, with a surface temperature around 6,000 degrees C, is a G-type star which falls into a G-O sub-class.

Colour is a fine indicator, and the massive hot stars — the O and B types — shine a hot greenish-white, while at the other end of the sequence the G and M types glow orange; and so on, through to the more uncommon deep reds of N and S stars.

The reasons for these differences are clear now that the internal processes of stars are better understood, and the life-cycle known.

It can be said simply, like this. The bigger the star the more it eats — the more it consumes, the brighter it shines . . . and the brighter it shines, the sooner it will dim out. Science has thus made stars predictable. The life expectancy of any sun is spelled out in the rate at which the atomic pumps at its core turn over.

Yet, this is not simply a tale of stellar doom, of nuclear furnaces burning themselves away to dark oblivion. This is a fascinating tale of creation. As science understands it, the tale goes like this:

Nuclear reactions are the essence of stellar life. Gravitational pressure, generation of heat, density, and the mass of material all play their part; but, in the main, it is the particles which comprise atoms that provide the potent forces which make stars shine.

We can see a newly formed young star within a dense cloud with concentric layers of differing conditions down to its core. There, at the highest pressure, heat, and atomic activity, the temperature can run from

Burning like a vast roman candle in space 5,000 light years from Earth, the "Cone Nebula" in the southern constellation of Monocerus is a prime example of star-birth. Metal-rich dust and gas clouds give the torch-like formation below a cocoon of fiery creation brilliant with the glow of emerging young suns. Point sources of infra-red and radiant signals suggest the possibility of systems of planets being formed with the infant stars and scientists specialising in the birth of the Solar System think they may be looking backward over 5,000-million years to a similar process to that which created the Sun's family.

(C.S.I.R.O. photo)

Remnant of a great sun. Wisps of an exploded giant star which erupted some 100,000 years ago and scattered its high-energy remains across the space of the southern Vela constellation. Named the Gum Nebula, after Colin Gum, the first Ph.D from Mt. Stromlo, this event in Vela left a legacy of the supernova — a Pulsar, known as PSR-0833-45 which was discovered by Australian radio-astronomers with the Mills Cross, near Canberra. The pulsar is located bottom centre of the picture — not at the energy centre of X-rays which is taken to mark the point of explosion. The neutron-star-pulsar, left over from the supernova, is believed to be an extremely dim object buried deep amid a crowd of small suns.

(Photo: Dr. Bart J. Bok — Schmidt Telescope, The Inter-American Observatory, Cerro Tololo Chile.)

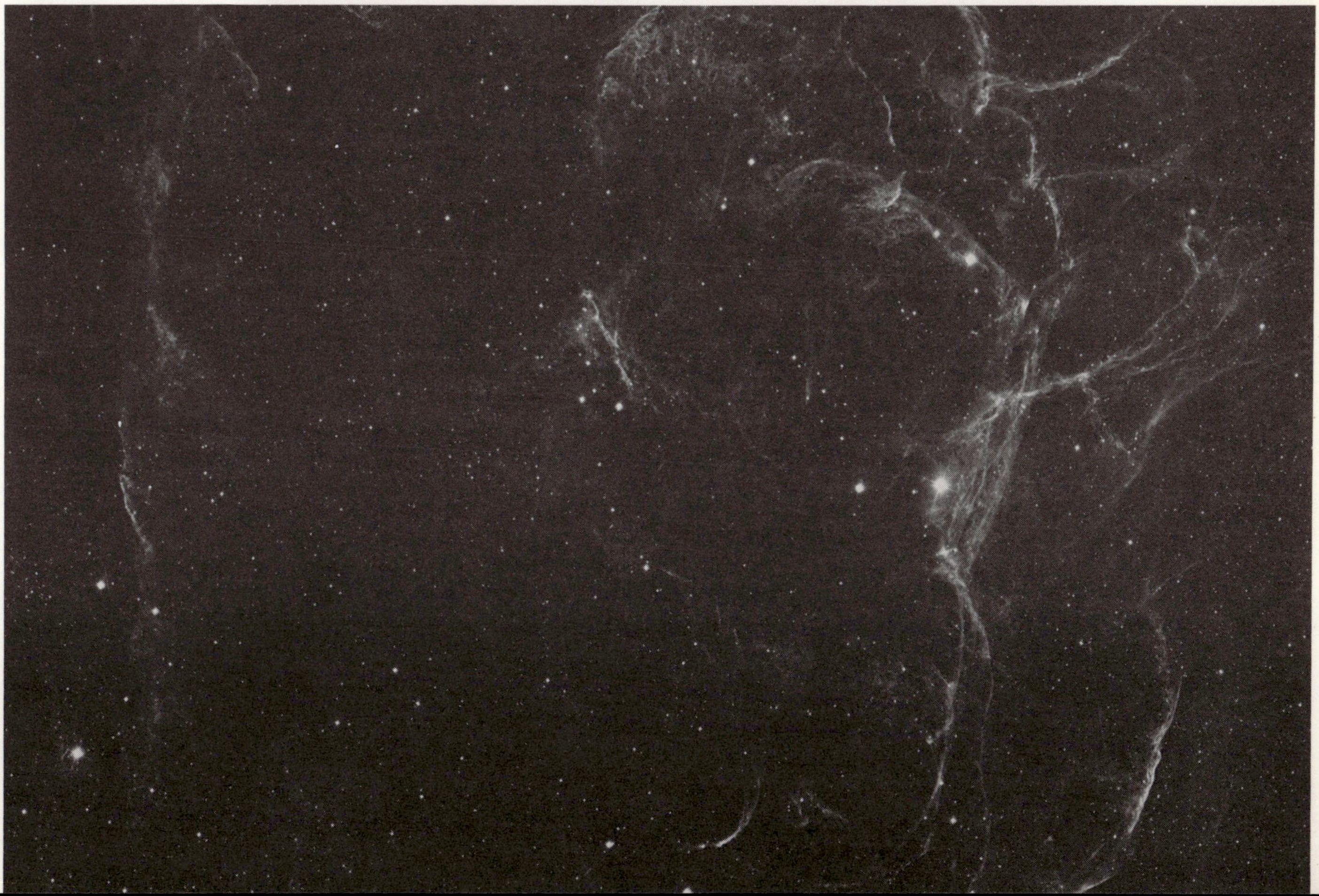

10-million degrees Celsius to as high as one billion! According to the size of star.

The process in this turbulent core has already been reviewed. Simple hydrogen is stripped — electron separated from proton — and refashioned, by collision, into two-part hydrogen (deuterium) which is fused into helium with radiant energy released in the process.

It does not stop there; for, eventually a time comes in the life-cycle of stars when the hydrogen is depleted to the point where it can no longer feed the core. The star then moves into processes leading to what is known as the "helium flash". The store of helium created during the star's lifetime is ignited. Other processes then take over. Helium is fused into heavier elements — to make nitrogen, oxygen and carbon atoms that are vital to forming the molecules which compose compounds essential to life-forming processes.

The star now presents a paradox. The core starts to shrink — but, since it creates greater energy, the outside of the star is expanded to release that heat; and the star moves into the stage of becoming a Red Giant. This is the point our Sun will reach in about 4-billion years from now. Stars of around that same size then go through their evolution to dwindle down into very dense pigmy stars which shine bleakly into space. They are called "white dwarfs"; their size has shrunk to about the same as Earth. They are dying stars, destined to fade into dark, cosmic cinders, probably to feed the dust clouds of the Galaxy.

Larger stars can have one of two fates. First, they can explode violently into "super-nova" and cast their processed materials of more complex atoms across wide space. As a side-show to this enormous firework display (see the Crab Nebula photo) they also create bizarre, shrunken objects of strange power and speed of rotation we call Pulsars. So fast is the action of the exploding star, the inner material of the core is crushed by "implosion". The very atomic structure is forced inwards and small stars of neutrons are created — the so-called Pulsars. These neutron-stars contain the most dense matter known in the universe. A piece the size of a marble would weigh thousands of tonnes. As well, the neutron star spins on its axis at a phenomenal rate. The one detected in the Crab Nebula turns 30 times a second — and signals its rate of rotation by throwing off excess energy in radio-waves which are detected by telescopes on Earth. Since these are radio pulses, and since such light as is visible from these stars also pulsates — they have been aptly named.

Pulsars have one other revealing feature; so extremely dense is their material — even though Pulsars could not measure more than 10 kilometres in diameter — their gravitation is fantastic! They draw material from surrounding space which, in these profoundly powerful circumstances, is excited into emitting x-rays. The single Pulsar in the Crab Nebula throws out energy in x-rays 10,000 times more powerful than the entire output of the Sun.

And the x-rays are clues to the second fate which may await even larger stars.

The greater stars will not explode, but will darken down into immensely compact black spheres of material. This material will be so dense it will keep on falling back into itself by the power of continually increasing gravitation. So solid will such a black body become, so gargantuan will be the ferocious gravita-

Solar systems in the making? Dark flinty chips of creation seen stark against the nebulous edges of gas clouds are thought to be the most likely candidates for that stage when gas and cosmic grains prepare to give birth to stars and planets. "They have not yet collapsed to the stage of forming a young protostar at their centre," remarks Dr. Bart Bok. These globules are being closely studied by radio-astronomers because of the variety of molecules in their tightly packed dust grains. This nebula is (IC 2944) in the Southern Milky Way; the smallest black fragment is bigger than the Solar System. The brilliant star is Lambda Centauri.

(Photo: Bart J. Bok — Cerro Tololo, Chile.)

tion, not even visible light will escape. These stars create sinks of gravity in the universe. They are a state of matter which has caught the imagination of scientists and public alike. They cannot be actually detected — except for the revealing x-rays. As with Pulsars, material drawn by the enormous magnetism of gravitation into this eternal graveyard of matter, throws off energetic x-rays . . . before it vanishes, into the "black holes" of space.

Stars behave differently because of their size and their position in the stellar life-cycle; astronomers know of situations where stars of the known same age, formed from the same original cloud, show a cross-section of type and size. These occur in the globular clusters.

Globular clusters are intriguing, engrossing objects for astronomers; a field of study in their own right. They each hold between 100,000 and upward of a million stars in single communities and they live apart from the rest of the Galaxy. Globular clusters orbit the heart of the Milky Way — but from far out. They are the original population of the "halo" and they follow paths widely erratic to the rest of the stars and gas. More than twice as old as our Sun, they are stellar antiquities, born when the great Galaxy was a sphere; created between 10 and 13-billion years ago from the pure, primal gas material. There was much less dust at that time.

The globular clusters are very low in the elements which stars fashion from simple hydrogen and helium; they are, as astronomers say, "metal poor". In this expression "metal" includes chemical elements from carbon up to the heaviest of the known minerals. Globular cluster stars have as little as 1/100th of the "metal" content of our Sun; and far, far less than much younger stars in the Galaxy disk.

There are about 200 of these ancient objects scattered through the outer "halo" of the Galaxy, at great distances. Some of them are 40,000 light years away. Three of the largest and the nearest globular clusters are in the Southern Universe — inaccessible to northern telescopes.

The most venerable, massive, and closest of these three clusters is the grouping of ten million suns in the constellation of Toucan . . . the object listed as 47 Tucanae. It is some 14,000 light years away and is probably as old as the Galaxy, upwards of 10,000-million years.

Believed to have formed in the first chapter of Milky Way history, when the Galaxy was a great ball of original gas, 47 Tucanae has hundreds of thousands of stars much bigger than our Sun — and the manner in which these have existed across the vast gulf of time has mystified astronomers until the present epoch of discovery.

With the advent of high-powered telescopes in the south the nearest and largest globular clusters are about to yield precious secrets. The first photographs taken of the 47 Tucanae cluster with the AAT 3.9-metre instrument on Siding Spring Mountain, in eastern Australia, have allowed observers to probe deeper into the nature of these close-packed suns than ever before (see photographs in this book).

The detail in such pictures, and the spectral lines (segmented light) these stars send to the Siding Spring instruments, will reveal the structure of the critical interstellar medium in these clusters. Thus, astronomers in the south will have a page of data from when the Galaxy came into existence.

These stars will yield secrets of conditions which existed more than 10,000-million years ago. That was when not only the galaxy, but the whole universe was very young — if accepted theory is right.

At the very least the tightly-packed globular clusters provide a perfect comparison with star birth regions close to the central disk and the inner spiral arms. The constellation of Monocerus, between Orion and Canis Minor, is rich in open clusters of very young stars and includes a curious object about 4,000 light years away which is obviously star-birth in progress (pictured); a wedge-shaped cloud of gas and dust, it is like a mighty roman candle burning into space. The young stars have the same hazy glow as the more famous Pleiades, where the energetic bright stars are still surrounded by gas and dust that fogs their true shapes (pictured).

The Pleiades in Taurus, and those bright, young, blue clusters in Hyades and along the gaseous disk are fast-burning types — big, young stars that will burn out their lives before they complete one rotation of the galaxy. And others will be born.

Out in the far reaches of the Galaxy's "halo" regions, however, the gas clouds and the brilliant young stars are missing. The star-making is ended; the gas has long since gone, depleted, and dragged by the grip of gravity into the flattened disk — perhaps to feed the enigmatic hidden heart of the Milky Way.

HEART OF THE MILKY WAY

The heart of the Milky Way Galaxy is masked from our view by colossal curtains of luminous gas interlaced with dark, deep, twisting lanes of stardust. The Galaxy's supreme splendour is hidden behind vast banks of swirling materials which are piled high beyond the glittering stars of the constellation of Sagittarius that swings directly overhead in the southern world from May to October.

The great Ferris wheel of stars, gas, planets, moons, revolves around the central point beyond the clouds, and, although we cannot view this magnificent scene, it is of great significance in our local creation. It is the focal point of galactic gravity, the axis on which the whole Milky Way turns. And more than that; thereabouts is the concentration of elements and the star-building material which fell back to the centre when the infant, bulbous Galaxy collapsed to its disk-shape, soon after formation. This heartland now bulges with gleaming banks of high-velocity incandescent gases, fleeing as though propelled by some enormous past explosion, and streaming among hundreds of millions of jostling suns which fill that inner space with incredible radiance.

No man has seen this grandeur. The scene is built from the glimpses which the probing fingers of modern technology have gathered. The immensity, the beauty, power, the fantastic light cannot be doubted, however; each discovery and each finding add to an awe-inspiring picture.

Within the regions of its heartland the Milky Way sponsors a fantastic level of star-making; the evidence now is that each single cubic parsec of volume is filled with gas many hundreds of thousands of times more dense than in our solar system; and that in this small volume of space there would be at least *one million stars!*

What does this mean in our experience?

Imagine a million brilliant stars crowding the sky around the Sun — filling our space *only two-thirds* of the way out to Alpha Centauri; imagine them shining fiercely — blue-white, green, yellow, red — through the thick, swirling gas clouds. It would be a scene alien to our eyes; and, it is calculated, despite the fogging by the gas clouds, the million suns would light the night sky with the brilliance of 200 full Moons.

It is the upheaval and convulsive energy streaming from such a concentration of stars and gas that has made construction of the scene possible by radio-astronomers. The picture has been built upon the electromagnetic radiation. The clouds behind Sagittarius are impervious to wavelengths of visible light, but not to shorter and longer wave-lengths in the spectrum.

The wavelengths of radio and infra-red thread their pathways through gas and dust grains. Thus, the radio pulse of the great heart of the Milky Way was detected in the first decade of radio-astronomy, when — in the early 1950's — Dutch and Australian pioneers in this new field plotted a source of emission known as Sagittarius A. Persistent and painstaking, they built their map of the radio sky, and with it grew their understanding of the physical and dynamic

Heart of the Milky Way Galaxy. The centre of the star-clustered gas clouds of the central disk of the Milky Way, close to the prominent star cluster (in the middle of the picture) is the closest view man has of the Hub of the Galaxy. The star density is so fantastically high hereabouts that astronomers view this region as a "window" into the star-crammed heart of our local star system. Each tiniest fleck of light is a sun, and the depth to which structures go show the scientists that the inner regions of the Milky Way contains billions of stars that we can never see.

processes which transmit signals across 10,000 parsecs of crowded space, to the listening antennas on Earth.

The radio-astronomers found that a whole variety of signals flow from the Milky Way. They came to recognise the radio fingerprints of stripped electrons of hydrogen caught in a spiralling magnetic field, called the "synchrotron effect"; they identified signals where electron and hydrogen proton were recombined; they learned the signatures of the life-cycles of the millions of stars — and, one by one, they came to know the specific radio patterns of atoms created by those stars.

This new radio "window" displayed a situation in the Galaxy's nucleus which stretched the imagination, which demanded incessant, and intense, observation.

By the early 1960's Sagittarius A was plotted on the graph and shown to be directly in line with the dynamical heart of the Milky Way.

Much more was soon learned about the Galaxy's radio heart. Sagittarius A was charted as an area of intense activity, about 35 light years in diameter, in which there are embedded several radio "hot spots". These are small regions, about one light year across, from which the energy emission is far more intense.

The picture which emerged raised new questions on the link between the central regions of the Galaxy and the nature of the spiral arms. What feeds the heart of the Milky Way to maintain the immense energy output? And, are the spiral arms replenished with gas streaming outward from the hub — or are they a left-over from the past and now are winding into the central area of the flat disk?

These questions became relevant when the charting of the central regions by radio disclosed two rapidly moving rings of gas on either side of the Milky Way's centre. Radio charts, fingerprint patterns, computer modelling, all told the tale this way: On our side of the galactic nucleus there is a belt of outward moving cloud, now some 3,000 parsecs from the dynamical centre, and travelling at 50 kilometres a second. This has been called the "10,000 light year arm"; and, if it maintains its velocity, it will carry fresh supplies of star-making gas out to the region of the Sun's orbit in about 300-million years.

There is a second gas arm on the other side of the central region, and, for some unexplained reason, this is moving out at almost three times the speed — at 135 kilometres a second.

Astronomy now has this firm evidence of gas flowing out from the centre — but, what does this tell us? What is the energy behind the outflow? Has there been some enormous explosion in the past? Are such explosions a regular feature which keep the spiral arms fed with new gas supply? If so — what is the nature of the explosions? In those incredible conditions, of hundreds of millions of stars jammed together like a vast globular cluster in the galactic heart, do giant stars explode more often than out in the "halo" regions? Or are there other kinds of explosions?

Explanation of these puzzles will offer deeper understanding on the physical and dynamic nature of the Milky Way which could be relevant across the whole universe; so, solutions are being eagerly sought.

Meanwhile, confirmation of the star-crowded aspects of the galactic centre came from two other developments. Infra-red astronomy opened a second window into the heart of the Milky Way, and telescopes found a "hole" through the gas clouds of Sagittarius, a narrow tunnel in the sky which gives a glimpse of a small corner of that clustered region. The first photographic plates used were highly sensitive to red light; and fascinated astronomers watched as millions of images of stars sharpened on their pictures — millions of stars jammed into the heart of the Galaxy!

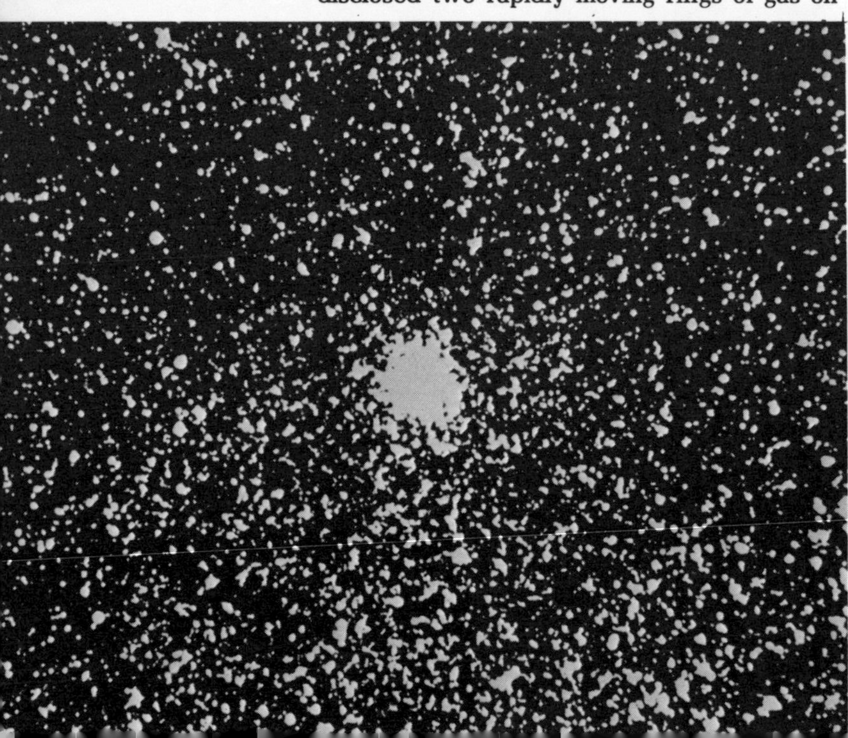

The view of the other side of the Milky Way Galaxy, looking through the Sagattarius "hole". The central feature is a massive globular cluster of upward of a million suns — but the most distant known. The cluster — NGC 6522 — is now estimated to be in excess of some 54,000 light years from the sun.

(Photo: Mt. Stromlo — 74-inch.)

Near the Heart. Behind the stars of the well-known constellation of Sagittarius are the close-banked star-clouds which hide the true heart of the Milky Way; this is a spectacular example of those clouds, the nebula known as Omega. This beautiful curving cloud curves about a central glowing starfield for hundreds of light years and shines brilliantly across a distance to Earth of around 6,000-light years . . . or 57,000-million-million kilometres!

(Photo: Mount Palomar — 200-inch.)

Another face of the Milky Way has also been unmasked. Astronomy has discovered that molecular chemistry is part of the great star system; and that production of substances which are keys to life-building processes on Earth takes place, on an astonishing scale, in the rolling clouds behind Sagittarius; indeed, deep in the regions surrounding the galactic heart.

Molecules which betray their existence with characteristic and identifiable outbursts of radio emission have opened a new avenue of research for galacto-chemists and those who seek answers to the questions men have asked through the ages . . . Are we alone in the universe? Is there other life out there — somewhere?

In the mid-1960's radio-telescopes in Australia locked on to a source of strong signals immediately overhead, in a cloud in the central disk which has been named Sagittarius B2. The length of the signals went down to a centimetre, and smaller, and were recognisable as the specific radio fingerprints of molecules associated with substances which build life-giving proteins and the nucleic acids.

The inference that the centre of the Galaxy could be a construction point for molecules of

Eta Carinae. The "elephant trunks" of the Southern Milky Way, seen in fine detail as never before; photographed by the powerful 3.9-metre AAT instrument on Siding Spring Mountain, the darker, narrow, twisted dust lanes known as "elephant trunks" are part of star creation.

This dramatic region is known as Eta Carinae from the Carina constellation, stars of which can be seen shining in the foreground. Located behind and between Southern Cross and Vela and Puppis constellations (see map) it is north of the bright Carina star, Canopus.

A galactic cluster. Named the Pendant, this cluster of varied stars shines against the active background of the Milky Way disk behind the constellation of Carina. Stars of different types and ages, their like are seen along the spiral arms of the Milky Way.

(Mount Stromlo.)

The Trifid Nebula. Part of the boiling mass of star gas and dust clouds behind Sagittarius which obscure the true heart of the Milky Way. This famous object has long been studied as a region where the star-making of the Milky Way still runs at an intense rate. Arms of darker dust clouds dissect the central region where young giant stars excite the surrounding gas and dust lanes into glowing nebulosity.

(Photo: Bart J. Bok — Arizona.)

chemicals essential to organised life sent a quiver of excitement through world radio-astronomy. It stimulated intensive searching, and evidence was quickly amassed on the scale and variety of this phenomenon.

An American study has found a ring of complex molecules circling the exact hub of the Galaxy, running almost all the way round, and about 1,000 light years out from the centre. It shows signs of being propelled outward at 100 kilometres a second.

In the southern sky, Australian astronomers have found other astonishing data. Small, dark, flinty-black globules in the vast Sagittarius clouds are packed with a surprising variety of different substances . . . from simple hydrogen and oxygen bonding into water vapour, up the scale to alcohol derivatives. One after the other, the chemical signatures came into the southern antennas; formaldehyde, cyanogen, carbon monoxide, water vapour, two-part hydrogen (deuterium) and, then, heavy water itself; others followed, such as hydrogen cyanide and cyanoacetylene and the methyls; until the list built into dozens. Even without telescopes fine enough to detect the still smaller signals at millimetre length, the patient galacto-chemists collected their evidence of the existence of methyl formate — the heaviest known molecule in the whole range. The next discovery was vinyl cyanide, a molecule with a feature which chemists call "double bonding", necessary for building into chains of substances.

Working with the giant radio-telescope at Parkes, New South Wales, Professor Ron Brown (a galacto-chemist at Monash University) has found that Sagittarius B2 has dense blocks of material where the activity in

The Black Clouds. One of the dense black clouds moving through a star field of the southern Milky Way which gave rise to Dr. Bok's phrase . . . "looking like holes in Heaven." The packed dark dust cloud blots out the stars behind. The so-called "hole" is only a hundred or so light years away.

(Photo: Bart J. Bok — Arizona.)

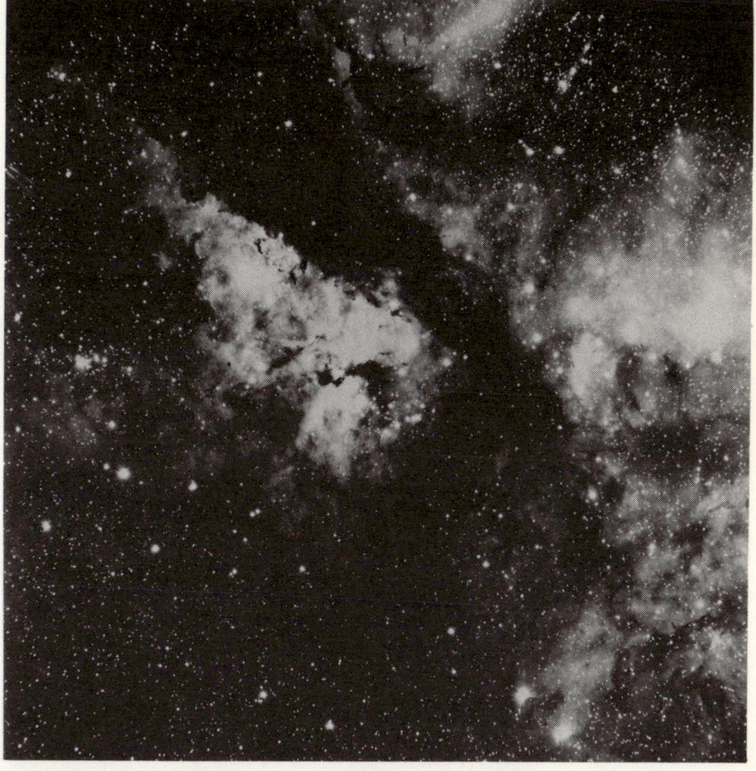

The Keyhole Nebula. This region lies close to the area in the Sagittarius starfields where astronomers can probe with red light to the other side of the Milky Way gas disk. Here the dark lanes of gas and dust grains give a deep sight into the structure of the galaxy, and so are known as a keyhole among the clouds.

signals is highest. He estimates the cloud has enough material to form upward of 100-million stars the size of our Sun — and that in the darker areas the molecular materials are predominant, not just occasional trivia. In the dark, flinty clouds, dust appears to be a minor fraction, but is considered essential for sheltering molecules from harmful short-wave radiation from stars — and for providing the grains on which molecules could form into longer chains.

He says, "We are confronted with a new situation in the sky. A few years back the idea would have been scorned, but, now, we must accept that life could have formed on Earth from such a cloud surrounding the planet in the early stages — after it had cooled. And there must be many, many places across the universe where such an event could take place. We would be very provincial, indeed, to believe now that life is not possible in other parts of the cosmos — that it is a miracle confined only to the little scrap of rock and water that we call Earth".

The search for greater detail is pushing ahead — and is backed by government funds.

Money has gone into building a small, but highly advanced radio-telescope, near Sydney, to work in the very short wavelengths required to scan the full face of the Galaxy's radio-heart. Local inventiveness makes this a superb instrument to meet the challenge. It is an opportunity similar to that given to major optical telescopes in the southern world; the critical areas of study come directly overhead, minimizing the distortions which face northern instruments looking down along the layers of the turbulent atmosphere.

As well, the new telescope has astounding capability by the use of a laser light beam — that is, coherent light which flows in tight patterns — and ultrasonics. These features will make possible a *spectral analysis of radio signals* — breaking them into revealing segments, just as optical astronomers do with light. This makes the new CSIRO 4-metre telescope at Epping, in New South Wales, unique in the world. What it will discover is still as unknown as the undetected range of shortwave signals which flood constantly from the heart of the Milky Way system.

Professor Ron Brown declares, "Yes, I believe life exists throughout the Milky Way Galaxy. There is every reason to accept the probability of living beings on many millions of planets where terrestial conditions are similar to Earth. There seems no question about that assumption in my understanding. Whether we shall ever be able to travel the vast distances or communicate by crossing space and time with radio messages is a very different question. The vast distance scale, the state of technology, the period of galactic evolution and history all intrude on this question.

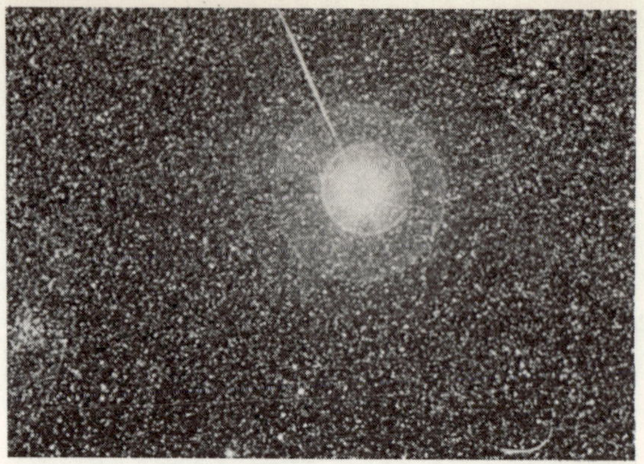

Blazing Sirius. . . most brilliant star in man's sky, 23 times brighter than the Sun and a mere eight light years distant. Photographed from Mount Stromlo (in hydrogen light) Sirius is clearly seen against the remote backdrop of myriads of suns lining the Milky Way. *(Photo: Mt. Stromlo)*

"I say the period of galactic history because some researchers into this matter put a limit of a few hundred years between the discovery of technology and the destruction of our species; and we need to take notice that a century or two is a very brief passage of time against the galactic clock which makes one revolution every 250-million years.

"Because of these factors I don't think it is likely that we are going to make any sudden dramatic discovery of some other civilisation in the Milky Way. There will be no such breakthrough. What is going to happen is that we shall press forward with our new-found knowledge and build on it until we have firm foundations for a good sound case that will argue how all this began".

In the future, however, the unravelling of the mysterious nature of the nucleus territory of our Milky Way system will hold many greater surprises. Some of these may come with the use of present scientific method, others by improvement of existing technology. Even more certain is that much valuable knowledge will be won from looking *outward* from the home Galaxy, across even vaster space to other star systems, some like the Milky Way, many wholly different, but all of them galaxies in their own right with a great deal to teach us about our own home in the universe.

LOVELY COMPANIONS TO THE MILKY WAY

High above the South Pole, out beyond the aged globular clusters in the "halo" of the Milky Way, and glowing against black depths of far space, are the two lovely Clouds of Magellan. Fuzzy white smudges to the unaided eye, they circle each other, and, together, revolve around the celestial pole in an endless cosmic *pas de deux*.

At declinations of 68 and 72 degrees south, they cannot be seen north of the equator; and, being unique in man's total sky, they are one of the main reasons for the new surge of interest, and the building of large telescopes in the southern hemisphere. For special reasons, astronomers now speak of the clouds of Magellan as "touchstones to modern astronomy."

They also have their place in history.

In Australia, legends have been handed down about the two white clouds; ancient Aborigines saw them as the souls of two sisters lost in The Dreaming, when the world was made. In the 15th century Portuguese sea captains reported to Prince Henry of two clouds, south of Africa, looking like "swans in heaven." Later, these two objects were named after Ferdinand Magellan who saw them on the way to his great exploration of the Pacific — and his murder in The Philippines in 1521. When the Clouds were named for Magellan, the Polish cleric, Nicholas Copernicus was penning the book which shook the foundation of the Christian Church with the findings that dethroned Earth as the Centre of the Universe and put the Sun in its place.

Long before this, however, philosophers in other civilisations had grappled with concepts far more expansive than the Earth-Sun beliefs which suffused European theological circles. The tantalising mysteries of the sky occupied minds in Asia, the Middle East . . . and the isolation of South and Central America.

High priests of the Maya civilisation fashioned their own remarkable astronomical methods; they plotted the motions of the Moon to within an accuracy of seconds, and deduced an oscillating universe with a cycle of birth, development, destruction and recreation over a period of 5125 years. It was a mighty achievement, for, as with all astronomy until the advent of the large telescopes, man lived under a closed canopy of the few thousand stars visible to the naked eye

In opening the mind of man to the greater universe, Isaac Newton made valuable contributions; one of these was the use of a curved mirror as a reflector in a telescope. Instead of the refractor system, which used lens and a long focal length to obtain an image, he collected light on a mirror and reflected this to a point of focus.

Observers began to notice wisps of glowing nebulous matter between the bright stars. These were called "nebulae" and were taken to be stray pieces of the material of the Milky Way — distant, but still part of one star system.

Ineffectual studies were made of these nebulae in the late 19th century. The American astronomer, Cleveland Abbe, wrote, in 1867, suggesting the Clouds of Magellan might be something far grander than extra pieces of the Milky Way, and studies were made of a few of the listed northern nebulae; but, unfortunately, these proved to be genuine fragments of gas and stars within the Milky Way. Thus, the 20th century dawned without resolution of the riddle.

Truth on the nature of the Clouds of Magellan and on many thousands of other small cloud wisps came slowly — almost unwillingly.

In 1912, in South Africa, the keen eyes of Henrietta Leavitt saw the first real clues. She

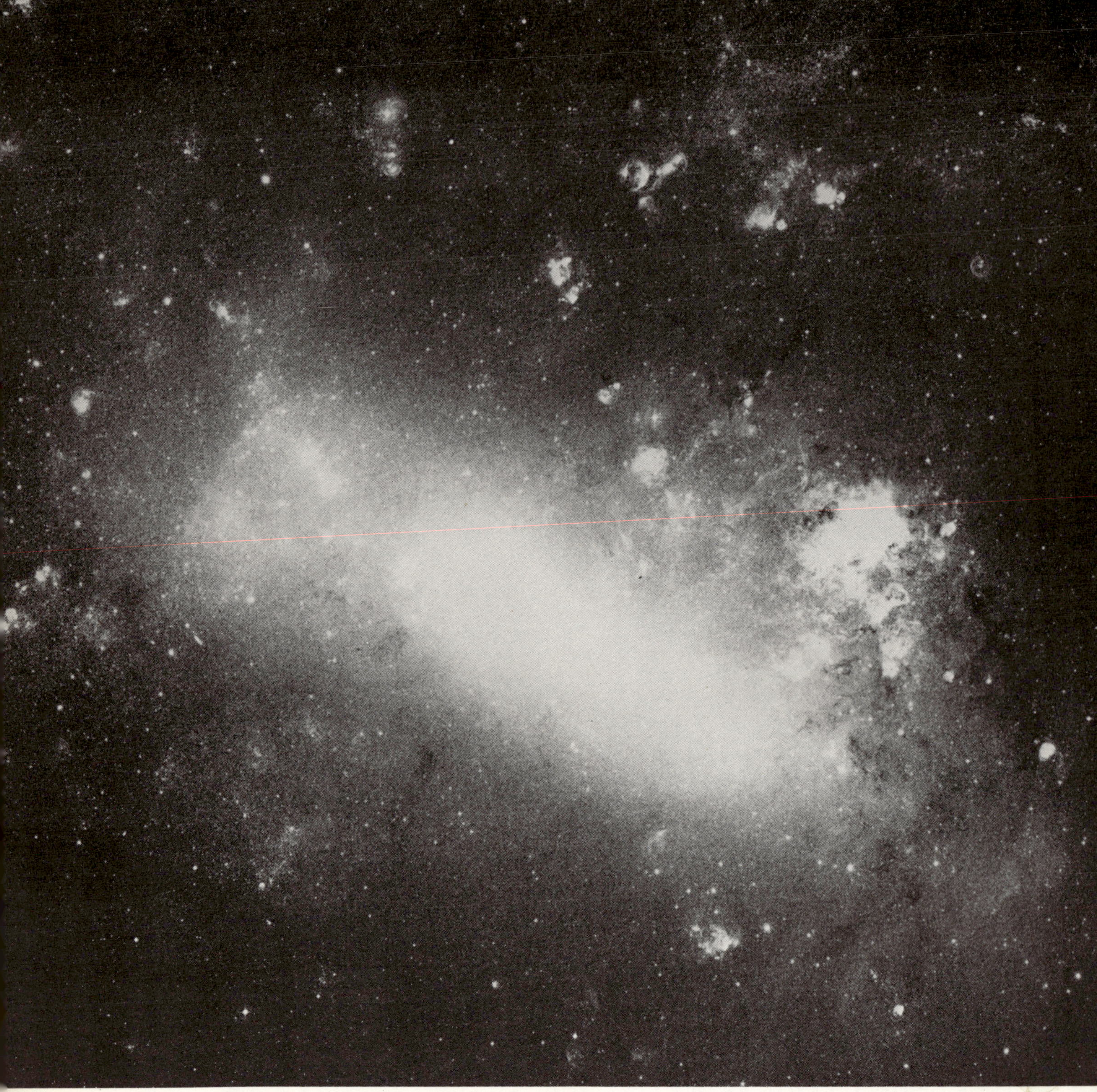

The large cloud in all its glory. This magnificent new photograph — taken with the big Schmidt camera-telescope at Siding Spring, Australia — gives unsurpassed detail of the stars and gas clouds in man's closest external galaxy. Creation and stellar evolution in another universe is here seen as it can be viewed nowhere else, because the Large Cloud of Magellan is ten times nearer than any galaxy visible in the northern hemisphere. Individual stars can be discerned out to the edge of this fine plate (specially sensitised for a 45-minute exposure) which covers an area of sky measuring 6.5 x 6.5 degrees.

(Photo: UK Science Research Council)

found more than two dozen of the variable Cepheid stars in the Small Cloud. The link she found in Cepheids, between the absolute brightness and the period of regular fluctuation gave astronomy the first true measure of the distance from Earth to the Clouds of Magellan.

Used as "space candles", the winking Cepheid stars broke down all doubt on the true nature of the two "lost souls" circling the South Pole. Far from being stray fragments of Milky Way material they emerge as complete, strikingly beautiful galaxies in their own right . . . and unique objects in several important aspects.

Locked in an embrace of common gravity, they are eternal companions of the Milky Way. Since they are so far away, their marvellous array of richly strewn star-fields, globular and open clusters and nebulae, can all be counted as being at the same distance from Earth . . . and that is a yardstick of wonderful value to the astronomer. Known distance is a great aid to charting stars, by magnitude and absolute brightness, and thus is a key to galactic evolution.

Now it is known, by virtue of Henrietta Leavitt and the use made of her "space candles", that the two fuzzy white clouds are so far away their light takes 160,000 years to reach our eyes. And, closely related as these two galaxies are, they are yet separated by 10,000 light years, an immense distance on our local scale. Each has already revealed many clues on the other; both have pages to add to the early history of our own galaxy, as well as to the over-all development of galaxies throughout the entire universe.

It is not just that the Large and Small Clouds of Magellan are totally without compare in our region of the universe. It is far more important that they are ten times closer to southern telescopes than any galaxy that can be seen from the north.

The magnificent Anglo-Australian Telescope erected on Siding Spring Mountain, near Coonabarabran in northern New South Wales is equipped — according to astronomical authority — with one of the finest reflecting mirrors ever made. This perfectly polished surface reflects stars in the Clouds of Magellan up to 100 times smaller than the famous Hale 200-inch (on Mount Palomar, California) can resolve in the nearest galaxy visible in the northern hemisphere; i.e. the great Andromeda Spiral, which is some 2-million light years away.

Detail of structure, content, behaviour of the stars in these beautifully close galaxies, make the two Clouds of Magellan virtual data banks of knowledge waiting to be tapped by inquiring astronomers. They are especially valuable because we can never see the entire structure of our home Galaxy. We can, with the two Clouds, however, look across 160,000 light years and actually witness *two complete galaxies* burn and evolve.

Nowhere else in the whole cosmic panorama does the universe lay out such an array of stellar riches so clearly to human gaze. Hydrogen cloud massing, star-birth, immense creation, supernovae — all are there to be photographed for study. With the power of great telescopes now appearing in the South, this part of the Southern Universe

One of the Family — This edgewise galaxy rides the sky in the southern constellation of Sculptor at a distance of some 6-million light years. It is believed to be a spiral formation similar to the Milky Way and though it lies beyond the boundary of the local cluster it is near enough and similar enough in shape to be counted as one of the local family. Its listed number is NGC55.

reveals stars as small as our Sun; we cannot see down to that scale anywhere else outside our Milky Way system.

Hitherto, the most powerful telescopes turned onto the Southern Universe were two 74-inch reflectors — one is at the Radcliffe Observatory, Pretoria, South Africa, and the other on Mount Stromlo, Canberra. Still with their limited power of resolution, astronomers have performed fine work with these two telescopes on the stellar treasures in the Clouds of Magellan; they laid the ground for the harvest about to be reaped. Atlases and star catalogues were painstakingly collated along with thousands of photographs.

This is the scene now being scanned with the great modern telescopes!

The Clouds of Magellan are wealthy with star-making material. Hydrogen gas flows through their structure in far greater amounts than the gas which feeds the spiral arms of the Milky Way. Consequently, there is motion and activity that is no longer present in most parts of our home Galaxy.

Globular clusters shine across 160,000 light years. With closely grouped millions of fresh, bright, giant suns, they are very "new" creations. Some of them are a mere 10-million years old — one thousand times younger than the aged stars of the globular clusters in the Milky Way's "halo"! Among these clusters are supergiants, fast-burning blue-white O and B stars, and extensive associations of suns, such as must once have existed in the young days of the Milky Way.

Some of these stars in the great clusters

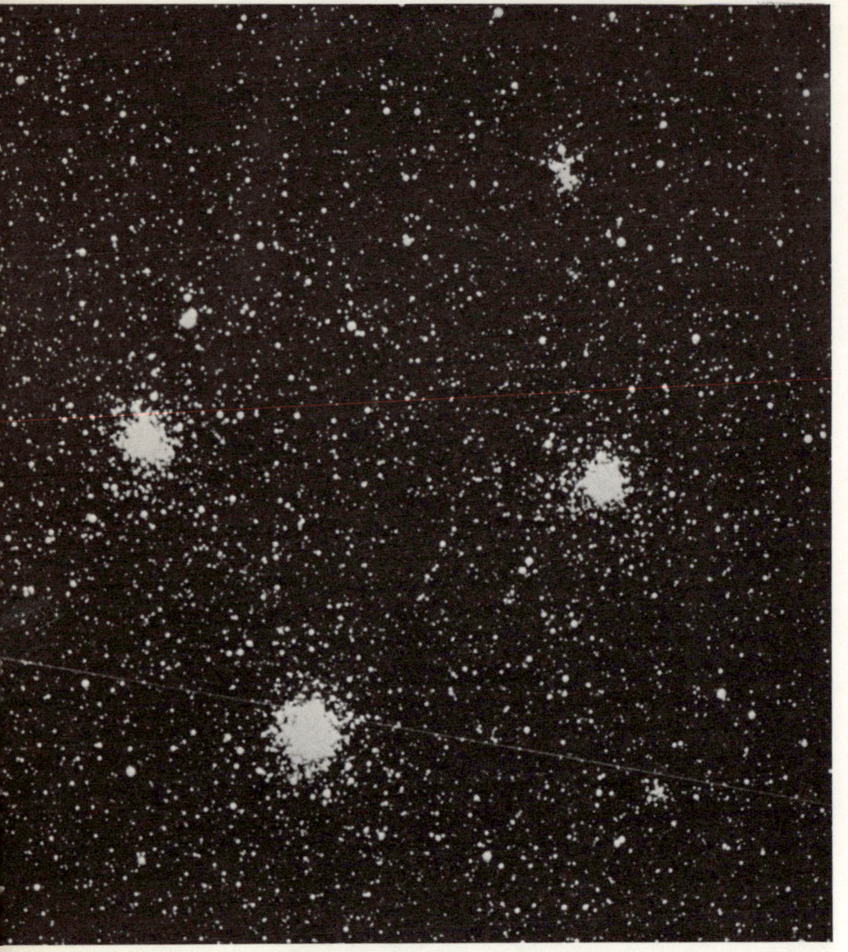

Young Clusters: Four colonies of stars in the Large Cloud of Magellan make perfect comparison with the oldest and nearest globular clusters in the southern sky. While the great 47 Tucanae cluster has stars known to be more ancient than 10,000-million years, these four young clusters in the Large Cloud are judged to be babies . . . a mere 30-million years old. It is still a mystery why such clusters can still form in the Large Cloud but not in our own galaxy. This opens a rich field of study for observers working in the Southern Universe.

The faint line running across this picture is the light of an Earth satellite, an unknown man-made object still orbiting this planet.

(Photo: Dr. K. Freeman)

The Great Tarantula Nebula: This magnificent photograph of the gleaming nebula in the Large Magellanic Cloud was taken by Professor S.C.B. Gascoigne with the giant reflector telescope at Siding Spring Mountain. It is our most detailed image of an immense star factory, a region which Professor Bart J. Bok says is one of the sky's best examples of a situation in which star-birth is now taking place.

This great nebula is already full and gleaming with the light of the type of giant white-blue O and B stars we see in our own sky. The excitement of the material which makes this vast conglomeration of gas and dust shine for 160,000 years across space to our telescopes is the same as that which we use in neon advertising. The Tarantula — officially known as 30 Doradus — is brighter than any advertising that man will ever be capable of attaining.

If this great nebula was placed in our own galaxy it would fill the entire area taken by the large Orion constellation — from Rigel up to Betelgeuse — and it would exude such brilliance we would never again know dark nights. It would be so bright it would cast shadows on Earth on moonless nights.

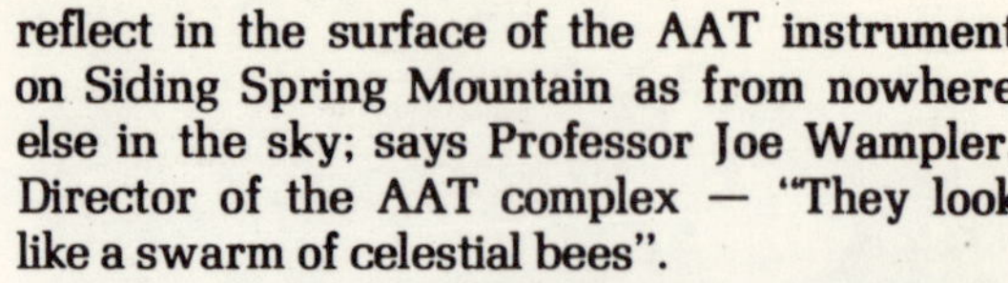

reflect in the surface of the AAT instrument on Siding Spring Mountain as from nowhere else in the sky; says Professor Joe Wampler, Director of the AAT complex — "They look like a swarm of celestial bees".

And star-birth; those regions where the cauldron of creation bubbles fiercest, have no compare in our sky. An outstanding feature of the Large Cloud of Magellan is the so-called Tarantula Nebula (pictured). Known to astronomers as Nebula 30 Doradus (from the southern star constellation) the object is one of the finest of examples of stellar birth. Embedded in clouds of hydrogen are hot, bright, young stars which cause the gas to glow like a brilliant advertising sign in space.

This great feature is so vast that were it place inside our own galaxy the Tarantula Nebula would fill the whole area of sky

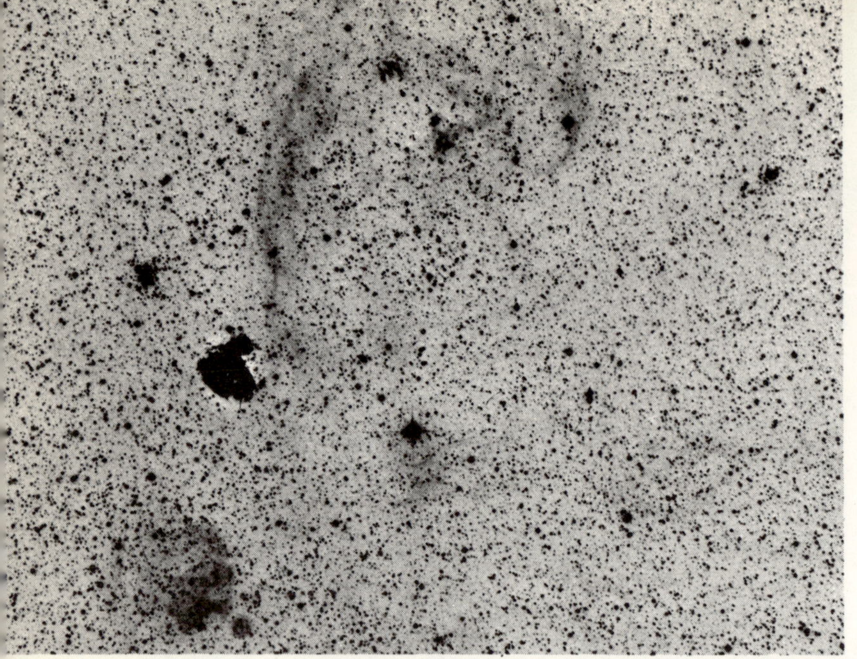

This exploding star in the Large Cloud of Magellan burst like a vast atomic bomb and spread its mushroom cloud of fabricated elements across a volume some 350 light years in diameter. When this light pattern left the Large Cloud the explosion was already 500 years old but the remnants of the star's photosphere and outer shells were still racing outward at high velocity. The object is listed as NGC49; again it shows no sign of a pulsar in its explosive centre but — strangely — the furthermost filaments of the expanding materials is a strong radio source.

(Photo: A.A.T. — Dr D. R. Mathewson)

Known as object NGC 86 the Fishing Net Nebula fills a volume 150 light years by 200 light years. Were it to have been placed where our Sun is when it exploded more than 200,000 years ago its materials would have been broadcast among stars within 100 light years of Earth. As it is — in point of real time — if we were able now to roam the space of the Magellanic Clouds the Fishing Net would no longer be visible.

(Photo: A.A.T.)

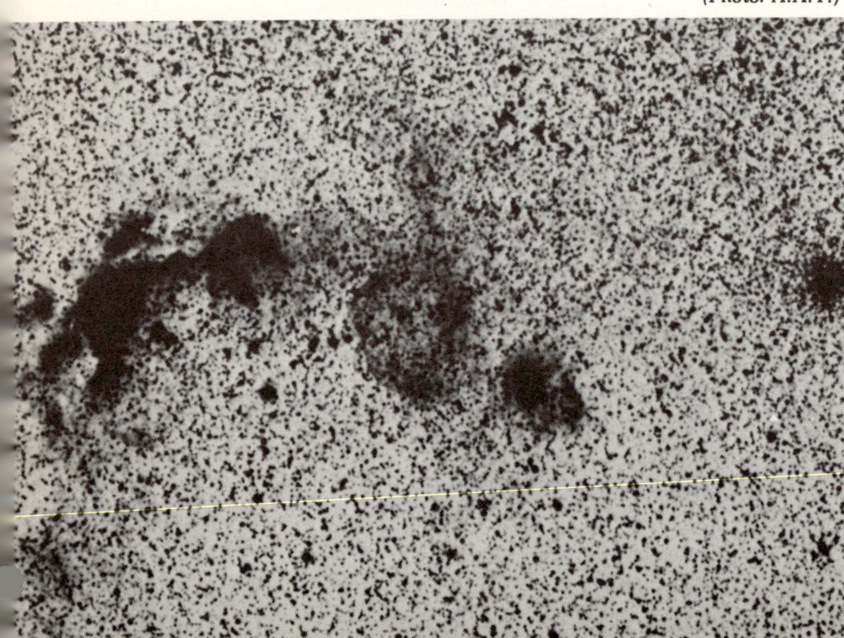

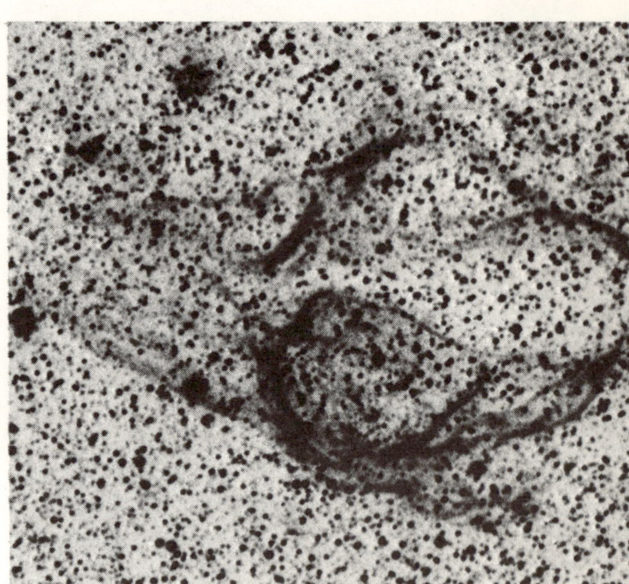

The Dinosaur Supernova. Among the first exploding stars to be discovered in an external galaxy — this enormous explosion of a giant star shows up against a background of a million young suns in the Large Magellanic Cloud.

Its discoverer — Dr. Don Mathewson of Mount Stromlo — sees it like a colossal dinosaur of expanding gas and dust, which chases its stellar prey, jaws apart. This picture was taken from Siding Spring Mountain with the Anglo-Australian Telescope. Such photographs — black on white — show up greater detail for astronomers to study.

which Orion now occupies. It would be so bright it would be visible in daylight and would cast shadows on Earth at night; the Moon would pale to insignificance.

In richly-strewn star-fields there are recognisable types of giant suns and numerous families of a class we cannot see in our own Galaxy. In some of the more crowded clusters there are hundreds of Henrietta Leavitt's Cepheids, winking out their distance with a mere three-day fluctuation period — a sure sign of being young and unstable (for reasons of physics we do not yet understand).

Among the stars in the Clouds are the expanding shrouds of exploded suns. These white scars of supernovae have long trailing arms of material, blown across light years by enormous eruptions. They result from explosions inside giant stars — thousands of years between each one — and are happening over colossal time (pictured). Fourteen of these explosions have been traced with modern telescopes; and they have much to tell of the life-cycles of stars in the two Clouds.

In these scenes, as in the so-called "emission nebulae" (the clouds of material where infant suns have thrown off their swaddling clothes of gas and dust) the evidence mounts that these are too young, forming galaxies. This is galactic evolution we are watching. There is star creation of a type and range which ended long ago in the home Galaxy; suggestions are made that the Clouds may be forming spiral arms of gas and stars: evidence indicates that each of these two star systems turns on an axis, as does the Milky Way.

All this builds exciting prospects for southern astronomy. The questions being asked are:

- *Why are Globular Clusters Still Forming in the Clouds of Magellan?*
- *Are the Two Clouds Much Younger Than The Milky Way?*
- *If so — What Process of Creation Brought Them Into Being? and When?*

Against general theories on the creation of the whole universe, answers to these problems may carry deep implications. The questions touch on the central theme that the galaxies were all born at the same time. A pioneer astronomer in the study of the Clouds of Magellan is Professor S.C.B. Gascoigne, of Mount Stromlo, the Commissioning Astronomer to the Anglo-Australian 3.9 metre Telescope. He says, "We do not have the answers, yet. We still must ask — Are the Clouds as old as our Galaxy? Were they created in the same way? And to answer we must work in greater detail, at greater depth. We have to measure star magnitudes, colors, and temperatures, as never before." The new, powerful telescopes have made this scale of exploration possible and it can only be done with the Clouds of Magellan. Compared to other galaxies, they are in our Cosmic backyard at a distance of 50,000 parsecs.

The strange theory of the "Black Holes"

Astronomers find strong sources of x-rays in the sky where no light shines.

In December, 1970, the world's first x-ray-seeking satellite was launched into orbit above the atmosphere from San Marcos Island, off the African coast. The instrument package was named "Uhuru" which is Swahili for "freedom". This remarkable experiment has since located many dozens of strong sources of x-rays, not only in our own Galaxy, but also in the Clouds of Magellan and other neighbouring galaxies.

Among "Uhuru" prizes were objects that switched off and on — regularly. Some of these were scanned by optical astronomers in Australia and America. Similar findings resulted. Workers at Mount Stromlo showed that companion stars revolved around an unseen object which was the emitter of x-rays; out of this the theory of the "black holes" in space was developed.

Expressed simply, this argues — with the backing of Einstein's theories on relativity — that if a star with more than twice the mass of our Sun explodes, the "imploding" forces can crush the structure of the sub-atomic neutrons — and go on crushing them into endless gravitational collapse.

For simpler understanding of these events, it can be said that the inward crushing force of the gravitation of this fantastically dense material is so fierce no light, no radio waves — nothing of the whole electromagnetic spectrum — can escape. The collapsing star, then, is a gravitational sink; x-rays are thrown off when matter is drawn into that sink, as they are with pulsars; but, once in, nothing ever emerges again.

Riches in the clouds

by
Professor S. C. B. Gascoigne, Mount Stromlo and Siding Spring Observatories
Commissioning Astronomer to the Anglo-Australian Telescope, Siding Spring Mountain, N.S.W.

The Clouds of Magellan are among the most important objects in the astronomers' sky. They hold great riches in supergiant stars, many wonderful clusters, remnants of supernovae, valuable variable stars, gas clouds thrown off by emerging young suns (emission nebula), and, because of this, present observatories in the Southern Hemisphere with superb observational opportunities.

As well, these two external galaxies are not visible to the observatories in the north; and they are also closer to us by a factor of ten than any galaxy that can be observed in the northern half of the world.

The proximity of the two Clouds has major advantages for southern astronomy. With the largest southern telescopes — the 150-inch reflectors in Australia and Chile — we can now measure brightness and colour of individual stars down to about the 23rd magnitude, which is about the intrinsic brightness of the Sun. In the northern world the nearest galaxy that can be observed is the Great Andromeda Spiral, some 2-million light years distant. This is so much further away from Earth that even the world's most powerful operative telescope, the Hale 200-inch on Mount Palomar, can only see down to stars that are about 100 times brighter than the Sun. The value of such detailed observation is very apparent.

Cousin to the Milky Way. The beautiful Great Spiral Galaxy of Andromeda is a giant outrider to the Milky Way and shares the role of dominating the local cluster of about 20 galaxies. Riding low into the southern sky it is the nearest major galaxy to telescopes in the northern world but at 2-million light years is still so distant its stars must be 100 times larger than those that can be seen in the Clouds of Magellan before the biggest telescopes can resolve them into detail. Two satellite galaxies ride with Andromeda, as the Clouds do with the Milky Way.

(Photo: Mount Palomar — 200-inch.)

It means that the two Clouds of Magellan are the only two galaxies in the whole universe within which we can examine the great clusters and resolve them into individual stars. The scale of colours and magnitudes which this offers is highly important to many branches of astronomy.

It is worthwhile fixing the scale of these two close companions of the Milky Way, and to do this I will describe their distance and size in kilo-parsecs: one kilo-parsec (1-kpc) is equal to 3,200-light years. Our Milky Way Galaxy has a diameter of roughly 30 kpc; our Sun lives 10 kpc from the Milky Way nucleus; the Clouds of Magellan lie some 60 kpc away.

The Large Cloud of Magellan has a diameter of 15 kpc (thus it is quite large, since its width would be equal to the distance from the edge of the Milky Way to the galactic nucleus, some 50,000 light years). The Small cloud measures about 10 kpc across, the distance at which the Sun rests from the heart of the Milky Way. A distance of some 10 kpc separates the two Clouds so that together they span an area of sky of 35 kpc, which is greater than the diameter of the entire Milky Way.

Although these two objects are separate galaxies they are closely related systems; radio-astronomers have shown how they are enclosed in a common envelope of neutral hydrogen and since this has been somewhat confirmed by optical observations we can consider the two clouds of Magellan as a physically connected system. This makes both clouds important as sources of various types of astronomical objects.

Examples of these are the valuable variable stars we call Cepheids and supernovae, or exploded stars. The Cepheids might be considered as the first great discovery made in the Clouds of Magellan. This came more than 60 years ago when Miss Henrietta Leavitt (of Harvard) noted the period-luminosity law for Cepheid variables while studying the two Clouds of Magellan. There are many hundreds of Cepheid stars in the Clouds with

fluctuation periods from 100 days down to 36 hours with brightnesses ranging from the 13th to the 18th magnitude. Miss Leavitt's classical discovery was that these stars could be used as standard distance indicators — or "space candles" — for external galaxies, as well as within our own Milky Way. She found the brightest Cepheids had the longest period of variability, which meant that if one Cepheid's period of fluctuation in luminosity was ten times longer than another then it was a Cepheid star that was ten times brighter. This gave a precision to the estimation of brightness in the Cepheid stars which had profound effect in measuring space and on our knowledge of the extra-galactic universe.

The advance in measurement which the hundreds of Cepheids in the two Clouds of Magellan offered has transformed their importance as celestial objects to among the two or three most critical in the sky. The Clouds are at a known distance and thus are sources of important data in various types of objects vital to understanding galactic evolution — a key to the whole universe.

Examples of these objects are the many supernovae discovered from Australia in the Clouds. These are the remnants of highly evolved stars which, toward the end of their lives, exploded with enormous violence ejecting great clouds of matter far into space where physical conditions are such that they are caused to radiate strongly in radio wavelengths. Because they become radio-sources and can also be seen by optical astronomers, the comparison of the two sets of observations can be used to reveal much about these important objects and the build-up of heavier elements throughout galaxies and the younger, metal-rich-stars. The Large Magellanic Cloud is now our most fertile field for the study of supernovae. Many other critical objects can be studied with similar intensity within these nearest two external galaxies.

Much knowledge has already been won. We know a great deal about their structure and their dynamics; for instance, we have found that in spite of its appearance of being spherical the Large Cloud of Magellan is really a flat rotating star-system only slightly inclined in facing in our direction. This allows us to make extensive calculations of its total mass and the distribution of its stars and gas and dust.

As well, we have discovered that the Small Cloud has stars that are far less rich in metals and heavier elements than stars in either the Large Cloud or in our own Galaxy.

These factors all combine to tell us things about the evolution of such galaxies and their history. Through this we shall come to wider understanding of the most important problem in astronomy — the knowledge of how galaxies in general are formed and how they evolve.

The two Clouds of Magellan have given us the opportunity and we have already gone a long way to finding keys to this evolutionary history.

The Large and Small Clouds of Magellan, seen through the outer-lying stars of the Milky Way — are two priceless objects of study for southern astronomers. Observers in the south can see stars down to one-hundreth of the size of any in external galaxies visible in the northern hemisphere. In the colour picture of the Small Cloud the closest globular cluster of stars to Earth — 47 Tucanae — is clearly seen.

(Photo: Vince Ford — Mount Stromlo.)

Giant on the mountain

The huge dome of the Anglo-Australian 3.9-metre (150-inch) reflector telescope looms above the landscape on the flat top of Siding Spring Mountain, in northern New South Wales. The great 500-tonne dome turns by electric motor-power to open its aperture to the part of sky astronomers seek to photography. This was the first major telescope to survey the southern heavens.

The flow of ideas, concepts, imagination are the life-blood of astronomy, but the bone and sinew of its current revolution has been superior technology. Nothing epitomises the skills of modern science, optics, engineering, cybernetics (computers), and mathematics more than the giant telescope now standing on Siding Spring Mountain, in the Warrumbungle Ranges, near Coonabarabran in northern New South Wales.

The Anglo-Australian Telescope was the first major optical instrument to be turned on to the Southern Universe. Astronomers who use the massive instrument say it is as good as any telescope yet constructed. The 3.9-metre reflecting mirror (150-inch diameter) is also claimed to be one of the two finest yet made; the other is for a telescope in Chile, the fabrication of which was encouraged by the success of the AAT's mirror.

The AAT is sited at Siding Spring Observatory which is the annexe to the national observatory at Mount Stromlo run by the Australian National University, and it came about from a joint agreement between the Australian and British governments, which was announced in April, 1967. The mammoth telescope cost $16-million to erect and equip, and six years to build. It was inaugurated by Prince Charles in October, 1974, and at once started its program of calibration and necessary adjustments.

The decision to build this great telescope in Australia arose from years of canvassing by astronomers who felt keenly that detailed and in-depth exploration of critical regions of the universe had been neglected. Among these campaigners was the former Director of Mount Stromlo, Professor Bart J. Bok, and his successor, the current Director and Professor of Astronomy at the ANU, Professor Olin J. Eggen. Weight was added to claims for modern facilities in the southern world by international leaders of astronomy, who made regular pilgrimages to use the then inadequate resources in the southern hemisphere.

The mobile part of this complex telescope weighs 350 tonnes and is guided to focus precisely on selected objects by a computer brain armed with powerful, silent motors. Telescopic equipment includes the capacity to divert starlight along various paths, for study, for intensification, for spectroscopy and for photography of various kinds. It also gives observers television pictures of the objects under study.

The beautiful mirror was cast in Ohio, U.S.A., from a material called Cervit which resists expansion from changing tempera-

tures. It was contoured and polished at Newcastle, England, by artisans of the Grubb Parson Company, and was so precise it gave reflections for fine photographs before the surface was finally aluminized with a brilliant metal coating.

This giant telescope is protected from weather by a huge dome, 120 feet above ground level, 150 feet in diameter, and weighing 500 tonnes. This massive canopy rotates on smooth bearings and is powered by servomotors so that the opening above the end of the telescope tube can be turned to any corner of the southern sky.

Nobody can forecast the working-life of such a telescope, but experienced astronomers talk readily of more than 50 years of useful work.

The resolving power of this giant is hard to define, since the main work of the reflector will be in looking out to extreme distances. Some experts, however, say the mirror will reflect the light of a safety match struck 3,000 miles away, and that it could detect a 40-watt electric light bulb as far off as the Moon. Its capacity to gather light is at least a quarter-of-a-million times greater than the best human eye . . . and, as well, it has the associated photographic equipment to capture and store that light for permanent record and analysis.

The two pictures span two centuries. In the illustration from the "Australasian Sketcher" of December, 1874, the astronomer is seen noting his observations on the transit of Venus across the face of the Sun — at noon, on December 9 that month. "Contact between the planet and the solar rim was clearly seen," he reported. Venus took 3 hours, 59 minutes, 28 seconds to transit the Sun. This result tallied with the same observation made from Tahiti a century earlier by Captain James Cook, the study which brought the navigator to the South Pacific and the discovery of the east coast of Australia.

The AAT 3.9-metre mirror reflector at Siding Spring has been described by British astronomers as the "most efficient telescope in the world". Its beautiful 16-tonne mirror is contoured to perfection approaching a millionth of a centimetre and is one of the two finest mirrors (and costliest) ever made. The other has gone to a similar telescope in Chile. The AAT controls are wonders of ingenuity and electronics. The sky co-ordinates are punched into a computer and the 350-tonnes move quietly, smoothly, and fix the target. The Commissioning Astronomer, Professor Ben Gascoigne, says it would register the light of a 40-watt bulb on the bright face of the Moon.

(National Library of Australia)

Cousin in Chile. The 150-inch reflector telescope at Cerro Tololo, Chile. Jointly owned by American universities it is virtually a repeat of the giant telescope on Siding Spring Mountain with the same kind of ceramic glass mirror. These two mirrors, say leading astronomers are the finest ever made, and they will spearhead the new era of exploration of southern skies.

(Photo: Prof. Bart J. Bok.)

The observer's perch is in the cage on top of the telescope's tube; from here a pin-hole through the centre of the photographic equipment allows sighting of the image reflected in the perfect mirror, some 30 feet below. Scientists are thus carried to the top of the open dome and ride, strapped in this seat for hours, while their photographic plates are exposed.

(Photo: Dr. R. R. Shobbrook.)

A COSMOS FULL OF ISLAND UNIVERSES

The Clouds of Magellan are portals to outer space. Their location — 50,000 parsecs above the South Pole — is our threshold to the greater universe, an arena so vast it is better described as — the cosmos. This is the great empyrean of the heavens, the region of supreme creation where scientific astronomy is joined by cosmology and modern physics in the search for understanding of the true Beginning.

Other smudges of haze dot the spaces between the stars of our Milky Way. In the big telescopes most of these resolve into other worlds, distant star systems of wonderful variety and form which fill black space to infinite distance, like an enormous swarm of fireflies on a dark night in the tropics.

Their forms are fascinating, with myriad shapes in every possible pose; spirals trailing loosely-bound filaments of light, spirals tightly wound; some on edge, some full frontal, others tilted and angled among billions of ellipticals — some of these with gleaming golden nuclei, like the yolks of vast eggs in the cosmos; millions of more galaxies irregular in shape or form, like the lovely Clouds of Magellan; some weaving glowing arms across the black backdrop, like the spidery antennae of insects.

All this vast, teeming cosmos of great galaxies, star-strewn worlds without count, was unknown to man until the middle of the present century.

Discovery of this immense cosmos had to wait on development of skills in optics, on the fashioning of the marvellous breed of modern telescopes; and on the method of establishing true distances with the standard "space candles" which Henrietta Leavitt discovered burning in the Clouds of Magellan, in the southern sky.

The existence of such a cosmos was not totally unsuspected. Discussion of the nature of the hazy smudges, called nebulae, aroused some speculation when the telescopes of the early 1800's revealed their variety of locations. By the middle of the last century, venturesome minds among a few philosophers teased at the idea that the Milky Way might not be the *entire* universe; that the nebulae could be worlds beyond worlds. The noted German thinker Immanuel Kant, gave them a scintillating title. He called them — "Island Universes"; galaxies floating in an endless ocean of space!

The idea of there being great aggregations of stars other than the Milky Way was too great a jump for the times; men talked a lot in private, but none would go into print and speculate on the true extent of the universe.

Some of these same questions were raised in Melbourne, in the 1860's, when the southern Australian state of Victoria contracted a brief fever for astronomical exploration. There had been much urging among learned circles in Britain for erecting a large telescope in "some part of Her Majesty's Dominions" — and civic fathers in Melbourne seized the opportunity.

What was to be known as The Great Melbourne Telescope was constructed in Britain and erected at the observatory in Williamstown. In 1869 it went to work on a study of the southern nebulae, and though it took beautiful pictures of the Moon, and was rated one of the world's biggest telescopes of its day, it missed the chance to resolve the spangled star-fields of the Clouds of Magellan.

It had a metal reflector which was a handicap. When the Melbourne Observatory closed its doors in 1946, the Great Telescope was shipped to Mount Stromlo where — with a new glass mirror of 50-inch diameter — it still does service.

In the lifetime of the Great Melbourne Telescope there came an astronomical revolution with a result as startling and historic as that which followed Galileo's first observation of the lunar mountains and Jupiter's Moons. It rose out of the discovery, from South Africa, of Cepheid stars in the Clouds of Magellan.

Using these stars cleverly — as standard "candles" for measuring distances — great astronomers (Hertzprung, Russell, Curtis, Harlow Shapley among others) laid the base for the final discovery of the great cosmos.

Argument and wrangling was still under way on the finite universe, when, in 1924, Edwin Hubble looked into infinity — with the brand new 100-inch telescope on Mount Wilson, California, then the most powerful instrument in the world. To his delighted gaze a northern sky nebula listed as M31* resolved into the glowing stars and gas, and he saw the yellow bulbous heartland of the great Andromeda Spiral — now considered to be very similar in appearance to our own Milky Way.

Hubble looked at another close galaxy, the visually attractive blue spiral in the northern Triangulum constellation — which holds the designation, M33 — and then at many others. A new epoch opened in human awareness of the universe.

In these great collations of creation Edwin Hubble saw recognisable stars, globular and open clusters, blue-white giants hanging fire on spiral arms — and the winking Cepheids, fluctuating just as they do in our own Galaxy; just as they do in the Clouds of Magellan.

Hubble was launched on the epic intellectual journey into the cosmos which has made his name immortal in astronomy.

He pushed human knowledge of the universe millions of light years into space, collecting evidence in light which had travelled 30-million years. He met difficulties, problems, set-backs, disappointment . . . and triumph. When he probed into regions too remote to photograph the revealing Cepheids, he used the comparative brightness of types of galaxies to judge far distance.

Then, in the precise manner of modern astronomy, he broke the light from distance galaxies into colour bands in a spectroscope — and discovered the Expanding Universe. Hubble rolled back the horizons of the cosmos with this momentuous finding. The billions of galaxies were great worlds in flight!

In every direction the conglomerations of stars and gas move away toward infinity and final oblivion.

So speedy is the recession of galaxies the red segments in the spectrum of the titanic outpouring of light from the tens of billions of suns are stretched out. The spectra received on Earth from these flying galaxies shifts heavily into the red end of the rainbow colours. The so-called Red Shift! This is the clue generally accepted as the key to the speed which such objects attain in their race outward toward the rim of our visible cosmos. How far the red light shifts became important because Hubble discovered that the further these galaxies are from our instruments on Earth — *the faster they recede*. This is the *Expanding Universe*! The deeper the distance, the further the light shifts to the red end of the spectrum. We now know of objects that transcend all past beliefs; galaxies that are receding at speed close to that of light itself. At that rate — at the distance implied — they become invisible to all our methods of detection.

"That is the point beyond which," says Professor Olin Eggen, "we shall never see."

Because of that, the territory over the rim of the observable universe is a non-scientific question, it can be only a matter of speculation. And it can give no answer to the question that surges in all our minds in contemplation of this picture — How did it all begin?

If this mighty concourse of creativity is flying outward — always outward — how is it impelled? What started it on its path? Those are prime, burning questions at the heart of modern astronomy and cosmology.

In 1936 Hubble provided a yardstick to measure the speeds of recession. He calculated that for each million light years of distance, a galaxy added a further 163 kilometres to its recessional speed . . . until finally it went out of sight by being so fast that light could not travel back.

The import of this calculation was that, traced back in time with such accelerating speeds, it could be shown that all the matter in the cosmos was in one single place at one time — some 5,000-million years previously. This finding gave nutrient to the roots of a concept expressed in 1926 by the Belgian

* "M" in front of numbers for objects stands for Messier, the name of a French observer who collated a catalogue of stars — and found these "nebulae" a worrisome nuisance that held up his task. Other objects bear the letters "NGC" (for National Geographic Catalogue). In later times, letters "C" and — rarely — "P" stand for the Cambridge catalogues, and one compiled of the radio sky from Parkes, Australia.

An edgewised galaxy. Companion Galaxy in our cosmic backyard is this tilted brilliant spiral — NGC 253 — in the constellation of Sculptor. Dark lanes of star-forming gas clearly mark its edges; it lies some 13 million light years distant.

(Photo: Mount Palomar — 100-inch.)

priest, Father Georges Lemaître. He had proposed the beginning of the creation of the cosmos as an enormous "big bang", the awesome explosion of all the material in existence in a great primeval fireball which fashioned all matter in its present form in the vast nuclear cataclysm and flung it all out across endless, empty space.

Mid-way through this present century man knew he lived in a cosmos overwhelmingly greater than had been realised by previous generations. He knew infinite space was dotted with "island universes", and that this whole structure was expanding outward at a speed that increased with distance. He did not know much more. There unrolled before the astronomers and cosmologists a complex research problem of colossal proportions.

It was the unravelling of total creation. Intrinsic to this daunting task was intense and detailed examination of the motions, structure, the nature and character of galaxies — and, indeed, of the stars and gas and dust clouds which form "island universes".

Hubble opened this work with a Herculean study of the various types of galaxies. He found a uniformity across the sky — a regularity that did not seem to disappear with distance. He classified galaxies as stars were classified — into divisions and sub-divisions. He arranged his parade into spirals, amorphous irregulars, and ellipticals. The spirals broke down into sub-divisions of tightly wound forms to "open" spirals, to "barred" spirals; he separated the various types of ellipticals from the near-spherical down to flattened varieties and the irregulars into segments of raggedness and colours.

All this showed a progression of characteristics — that most spirals are highly flattened and rotate quickly with considerable star-formation taking place; that the ellipticals give little sign of star-making and rotate slowly — if at all. There were further valuable clues. The laws of motions for celestial bodies illustrate that galaxies generally live in clumps — galactic gangs which (like stars) astronomers call clusters. From this finding mathematical channels were opened to calculate the total masses of these galaxies.

This factor is important to probing the nature and behaviour of the whole cosmos, and so it became even more urgent to understand how individual galaxies were structured, what they were composed of in terms of stars, gas, dust, energy, and those elements which stars build from the original building-block atoms of simple hydrogen.

Attention fell on the inner life of the Clouds of Magellan — and on the other members of the galactic tribe inhabiting nearby cosmic space . . . the so-called Local Group.

So far as can be seen there are about twenty member galaxies in our local cluster; there may be others in that part of the cosmos hidden by the cloud banks behind the Sagittarius and Carina constellations. We just don't know. Of those we do know, three dominate the scene — the spiral Milky Way, the comparable Andromeda Spiral (which barely shows its nose into the southern sky) and the blue, open spiral M33, in the northern Triangulum. These three massive and brilliant spiral galaxies reside within about 2-million light years of each other, within a total volume of 3-million light years for the overall region of the Local Group. In that vast volume are ten elliptical and seven irregular types of galaxies — two of the latter a mere step out in the southern sky from the nearby Clouds of Magellan.

These two small galaxies — sometimes called dwarfs — were only found by accident from South Africa in the mid-1930's; but, though small, they are valuable to southern observers since they provide classic samples to compare with the Clouds of Magellan.

They are the next objects out from the Clouds; one in the constellation of Sculptor is about the same distance again from Earth as the Clouds, and the other — in Fornax — lies another 300,000 light years beyond the Sculptor irregular. They are not cosmic prizes in themselves. There is little glory in our view of these two unpretentious members of the local galactic family; the colour arrays and the magnitude studies of these star systems from Australia and South Africa show them to be similar to stars in the aged globular clusters in the Milky Way "halo" regions.

What does this tell us? If the signs are read correctly, it says the Sculptor and Fornax galaxies are systems in which the business of star-making has ground to a near-halt. They show all the signs of senility; and so they are a priceless comparison for southern astronomers to match against the bubbling cauldrons of stellar evolution now clearly displayed in the nearer Clouds of Magellan. Indeed, some observers estimate that the star-pots of the Sculptor and Fornax systems went off the boil some thousands of millions of years ago. And to know why, in a universe where creation is supposed to have occurred at the one instant, is intriguing to cosmology.

A local member. This lovely spinning galaxy is in the sky above the equator — in Triangulum — but it is a major star-system in our local cluster of galaxies and from 2-million light years shows the valuable Cepheids which serve as "space candles" to measure deep distance. It is NGC 598.

(Photo: Mount Stromlo — 200-inch.)

Photographed with the giant telescope on Siding Spring Mountain, this lovely spiral galaxy graces the Southern Sky near the south celestial pole. It has no name, only a number — NGC 1566 — and is a spiral universe which may either be winding up or spreading outward. The light from this galaxy tells us it lies about 100 million light years away, but it is the nucleus which is fascinating Australian astronomers. The central region shows the liberation of enormous amounts of energy. From Siding Spring the great telescope has seen violent chaotic motions of gas and dust with velocities of upward of 2,000-kilometres per second in the glowing nucleus. This is an explosive process not yet understood which may have direct relevance to the suspected explosions at the heart of our own Milky Way Galaxy.

This brilliant galaxy is 20-million light years from the Milky Way — not a member of the local cluster but of direct interest. This object is known as a "barred" spiral and research seeks to show whether we here witness a stage in the evolution of one type of galaxy to another. The centre is strongly packed with material but the arms do not (yet) trail out so far into space.

This stupendously energetic galaxy is in the constellation of Centaurus and has the catalogue number of NGC 5128. It is a strong source of radio emission and is remarkable among types of external galaxies in that for many years it was considered an example of two galaxies colliding. More detailed pictures show its circular form to be sub-divided by belts of gas and dust tens of thousands of light years thick. Unknown to man until its discovery in 1827, this great galaxy is now seen to contain billions of white, blue and red suns. It is about 15-million light years away from Earth and receding at a speed in excess of 1,000 kilometres a second.

A Cousin to the Milky Way — From 30-million light years away this beautiful spiral galaxy sends its beams across the band of the Milky Way to Earth. A southern sky object needing clear vision for study, it is ranked among the larger spiral galaxies in the sky and very similar to the Milky Way in size and appearance. Listed in the catalogues as NGC 2997, it is among the least studied of galaxies. It rides the blackness of the universe out beyond the little-known constellation of Antila, between Southern Cross and Vela.

In these dwarf ellipticals — and in others ranging far out to depths where stellar resolution is possible — the big telescopes have not found a single blue-white supergiant, of the types burning along the spiral arms of the Milky Way, and in the nebulae cradles of the Clouds of Magellan. Their absence is a classic sign of old age in the star system, since these are the bright young giants that gulp their substance so quickly their life-span is less than a single galactic year.

When the star-lamps burn low it means the gas supply is being shut down. It means, too, these unpretentious objects, in their heyday, were glorious spectacles of bright blue giants; and that all galaxies, whether they rotate as fast as spirals, or as slowly as ellipticals, and irregulars, come to a time when the fuel runs out; unless, of course, it is shown that there are limitless supplies of gas floating in the spaces between the galaxies, which some can gather in to keep their star-making pots boiling.

Once the existence and nature of the cosmos was discerned, this unseen intergalatic medium at once assumed importance; for, not only could it be an extra fuel to keep the stellar evolutionary cauldrons bubbling in those galaxies able to attract fresh supply of fuel, it could possibly have effect as a medium of gravity in slowing down the outward rush of the fleeing galaxies. Cosmologists had to ask — Is there a slowing down in the speed of recession? Will all this matter reach a point where it starts to return under the pull of gravity to the point of the original explosion? Is the old Mayan concept of an oscillating cosmos near the truth?

Even as these very questions were being hammered into forms in which they might be investigated, science and technology made forward leaps which uncovered unsuspected aspects of the greater universe. Horizons were pushed still further out, time was rolled back to enormous depth, and an even stranger cosmos was revealed, one of violence and enormous energy.

Astronomy and cosmology came to another plateau; they came face-to-face with an invisible universe. Radio-astronomy and the Space Age entered the great arena.

Galaxies in collision. Of great interest to southern astronomers these two inter-acting galaxies — object NGC 2207 — in the direction of Sirius, show features in this large telescope picture not seen before. Tidal motions of intergalactic gas can be seen, and these are believed to be due to gravitational struggle between the two approaching island universes.

(Photo: Dr. R. R. Shobbrook — A.A.T.)

Burning like cosmic lamps in the constellation of Dorado — in the general direction, but well beyond, the Clouds of Magellan — these are two typical elliptical galaxies. The oval one is NGC 1553 and the more circular NGC 1549. They show none of the graceful form of spirals and every sign that star-making is close to an end. Peopled mainly by old stars they nevertheless are striking objects of great luminosity. These two are the brightest of a small group of galaxies resting at some 36-million light years distant.

(Photo: Dr. R. R. Shobbrook.)

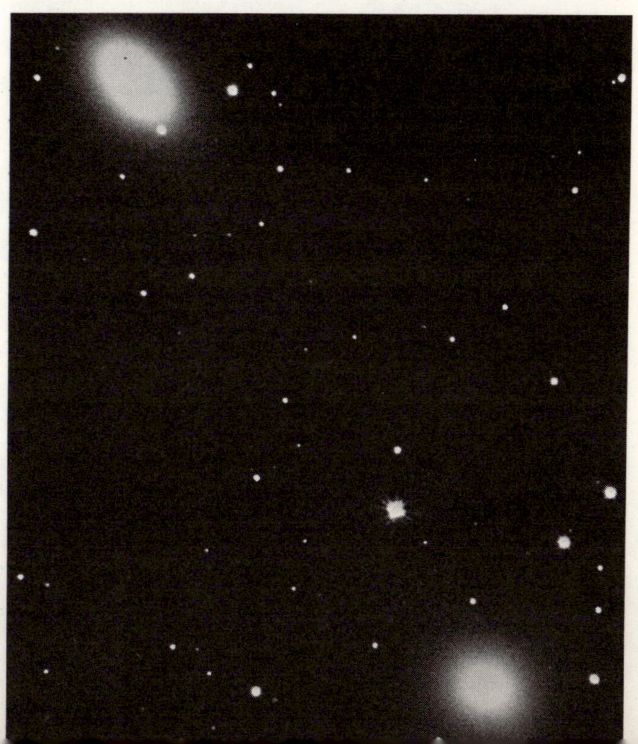

AN INVISIBLE UNIVERSE IS UNVEILED

By the early 1950's the probing optical astronomers were launched on the Homeric intellectual journey which unfolded the awesome vista of a limitless cosmos. They discovered space to be unfathomable, and that light from the outermost objects travelled for thousands of millions of years. The faint glimmering of distant galaxies thus came from nuclear fires which were burning long before the Sun and its planets existed.

As the second half of this century opened, men were looking back to the very Beginning of creation. Exploration in these years was pushed to the limit of the world's great optical telescopes. With these great instruments the astronomers scanned the architecture of the universe; they counted the fleeing galaxies in their billions, and their teeming populations of suns in *billions of billions*.

The wonders of the star-worlds now stood revealed; through the reflecting eyes of the telescopes — which "saw" no more than human eyes. Understanding of the greater universe was all based on tales told by light

Charting the Radio Universe. With such instruments as the great Parkes radio-telescope (foreground) and smaller dishes (which can be used as baselines to measure objects in the universe) the face of the invisible cosmos was unveiled. This magnificent radio-telescope has charted the radio-sky of the southern universe, located the first quasar, and among other things took part in the Apollo moon landings — including the space rescue of the Apollo 13 mission in 1970 when the lives of astronauts were threatened by malfunction near entry into orbit round the Moon.

— the narrow middle band of the wide electro-magnetic spectrum.

Unregistered in man's learning was the invisible face of the cosmos. The short-wave radiation of gamma rays, x-rays, ultraviolet could not penetrate Earth's magnetic blanket; and evolution has not equipped human beings with senses to detect the wavelengths of infra-red and radio.

That situation was already being corrected, however. Even as the great Hale telescope moved into its years of historic discovery, technology was reaching beyond the limiting atmosphere to gather the unseen light of infra-red radiation and the longer wavelengths of radio signals. Human intellect compensated for the omissions of evolution. New "windows" opened on the cosmos; astronomy broadened its view and opened its most exciting and revealing epoch.

Radio-astronomy blossomed from war-time experience of the development of radar. British and Australian pioneers in this field devised instrumentation to detect specific wavelengths bounced back from attacking aircraft; as well, they found that "background" static from the sky was a nuisance in their search for accuracy.

It was known that this constant radio "whispering" came from the Galaxy. In the early 1930's, Karl Jansky, a Bell Telephone worker in the United States, had rigged a 20-metre aerial to identify worrisome static as radio emission from the Milky Way; and Grote Reber, in Chicago, in 1937, erected a metal "ear" in his garden which charted several areas of intense radio static from the galactic centre.

After the war, Dutch astronomers at Leiden, and ex-radar workers with the division of radio-physics of the CSIRO (Commonwealth Scientific and Industrial Research Organisation) in Sydney, developed installations to explore the mystery of the "whispering" universe. Workers in Australia quickly identified one source of radio waves as being the Sun, but this was only a first faltering step.

A former radar operator, John Bolton, constructed an aerial on a cliff-top at Dover Heights, Sydney (pictured) and in a ramshackle old shed patiently built detection equipment based on his radar experience. The aerial looked like a large iron bedstead; the receiving equipment was crude by today's standards. Yet, it opened a new "window" that was to reveal an invisible universe of incredible power and turbulence.

The historic Australian 'Bedstead' Antenna. This remarkable device made the first detection of separate signals from radio sources in the universe. The pioneer Australian radio-astronomer, John Bolton and colleagues designed and built this antenna on a cliff-top above the Pacific Ocean, at Dover Heights, east of the city of Sydney. Later to become an historic monument of science and known as Bolton's 'Bedstead' aerial, it was put to work in 1949 and soon located the first six cosmic radio sources. Among these was detection of radio waves from the Sun, from the Crab Nebula and a star then known as 3C48, afterwards shown to be a powerful radio-emitting quasar. This latter object is more than 1,000-million light years from Earth and was the first radio source detected by man from such a vast distance.

The two antennae shown in this historic 1949 photo used an interferometry technique later copied across the world. The 'Bedstead' aerial detected radio signals from cosmic objects as the spinning Earth brought them above the horizon; the other antenna (dish) recorded radio waves from the same object after they had bounced at an angle from the surface of the Pacific Ocean. Measurement of differences in these two signals gave precise co-ordinates so that the sources of radio waves could be optically identified.

From this same site in the early 1950's Bolton and his colleagues also identified the first radio signals detected from the centre of the Milky Way Galaxy.

(C.S.I.R.O. photo)

Bolton believed the constant hissing and whispering in the sky could be resolved into discrete (separate) sources; his equipment was built to prove this assumption.

By mid-1949, he isolated distinct and separate radio points in the southern sky which he could identify with known optical objects. One of these was the Crab Nebula (pictured) which is the expanding remnants of a star that exploded in 1054AD. Chinese astronomers recorded that event and said it was so bright it was visible in daylight. Bolton found it to be a strong broadcasting station in the heavens. Among other points in the sky was another powerful radio source, an object listed as a star with the number 3C48 in the Cambridge catalogues.

Why should a seemingly unimportant, fuzzy, distant star throw off such enormous energy? When Bolton wrote a report of this work for an international science journal the answer to that question was not known. The words of his report, however, contained clues that helped to unveil the invisible universe.

A radio-telescope is comparable to a large and elaborate radio receiver. It collects the longer wavelengths of the normal spectrum and converts them to information which is registered in graphs and stored in computers. It deals with the same kind of waves as do normal radio receivers, although those from space run over a wide range of length — from the usual millimetre or centimetre waves to exceptionally long waves beyond ten metres. Various antennae were developed to catch differing types of wavelengths, and, as the work of exploring the radio universe developed, these became more and more sophisticated — as did the electronic equipment which dealt with the data flooding from the cosmos into the new "ears" of astronomy.

This technical advance was bound into the main stream of astronomy by a classic discovery made by a young astronomer at the Dutch radio-astronomy department of Leiden University; he predicted there should be a radio emission from the so-called neutral (cold, rarefied) hydrogen gas in the Galaxy with a wavelength of about 21 centimetres. It was a key to the future. Eventual observation by two American workers turned the key and opened the radio "window" to its full extent.

Lunar occultation of a radio source can provide extremely precise knowledge of its position and structure. The curve shows how the radio signal from the two-component quasi-stellar source 3C273 was blacked out by the Moon on August 5, 1962, as recorded by Australian workers using the 210-foot radio telescope at Parkes, near Sydney. The step in the descending curve (left) indicates where component B started to disappear. Because of their orientation with respect to the moon both components emerged together.

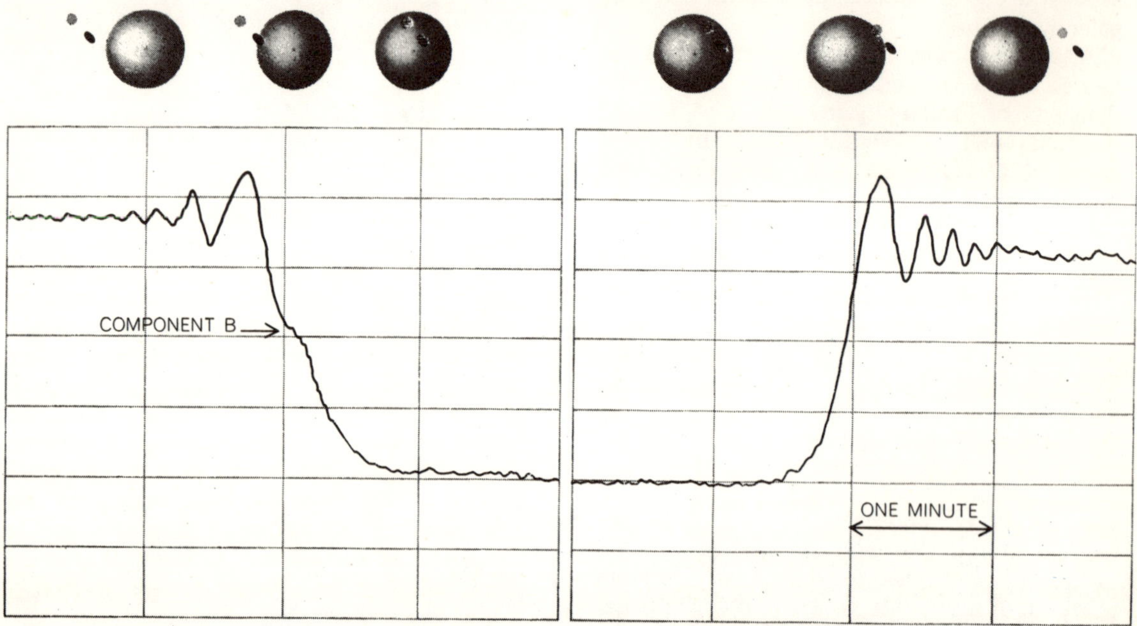

Stars might be radio mutes — but the clouds of gas and dust in the Galaxy, and out in the cosmos (that was to be highly important), signalled strongly with radio waves from the frantic electrons whirling about in their masses.

Radio waves travelled through such clouds and often re-emerged carrying added signals that were fingerprints of the variety of atoms in the clouds. This meant radio-telescopes could "see" into cosmic corners hidden to the big optical mirrors. The radio waves from the gas clouds, and educated guesswork, made possible the exploration of the true spiral shape of the Milky Way — and of the heart of the Galaxy with its closely packed giant suns and high-energy cloud masses.

By the early 1960's, the stage was set for stunning discoveries that were to send ripples of excitement through the world of astronomy. Consider these events:

An investigating rocket launched from the American White Sands missile range on July 18, 1962, carried sensors that detected a strong outpouring of x-rays from a source in

The visible pulsar in the Crab Nebula has winked out its existence to optical astronomers — proof of the phenomenon found by radio. Two stars — arrowed in the Crab — are seen in the top picture; the Pulsar has winked out in the second picture. The time for rotation of the pulsar is 30 times a second, calling for high speed pictures of 1-60th of a second between these two events. The Crab Pulsar is tremendously heavy, but in size very small. Its diameter would not reach more than four or five kilometres.

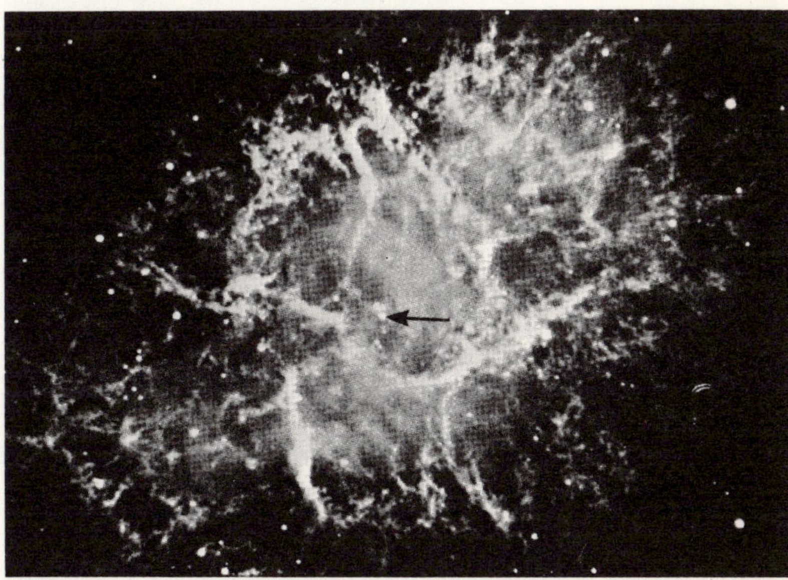

The exploding Crab Nebula erupted in our view in 1054 AD; noted by Chinese astronomers it was so bright it was visible in daylight. Lying 6,000 light years from Earth it draws much attention as an intriguing object with clues to other happenings in the universe — and because the fastest rotating pulsar yet discovered lies in the central region (arrow).

These are high resolution pictures of the pulsar in the Crab; taken through a rotating shutter — by the Director of the Anglo-Australian Telescope at Siding Spring (when he was at Lick Observatory) Professor Joseph Wampler — they show how the Crab pulsar blinks out some 30 times a second. This little pale blue star emits x-rays with energy 10,000 times the total output of the Sun, is a few kilometres in diameter and is composed of sub-nuclear material that is so dense a cubic inch would weigh 10-million tonnes on Earth.

the southern constellation of Scorpius. The source was named Sco-X-1. It gave birth to x-ray astronomy. Refined equipment was flown by rockets and big balloons from America and from Australia — in Adelaide, Tasmania, and from the Woomera ranges.

The list of X-ray sources in the sky exploded into more than one hundred — rapidly. One of them was the Crab Nebula. Originally found, to be emitting radio waves (by the bedstead aerial at Dover Heights), this astonishing object poured out x-rays with a total energy 10,000 times greater than the entire output of the Sun! Here was yet another mystery.

Optical astronomers joined eagerly in this new inquiry; the Sco-X-1 x-ray emitter was seen to be a small, faint, blue star. The big mirrors turned on to the active x-ray region of the Crab Nebula — proven as a star which had exploded some 900 years earlier. The x-ray quarry proved to be an elusive glimmer of pale light, a fragment of star-fire that actually blinked rapidly into the staring eyes of unbelieving astronomers (pictured). Mystery piled on mystery — until future developments brought forth the solution.

August 5th, 1962; — and in a former sheep paddock at Parkes, 480 kms northwest of Sydney, the most powerful radio-telescope in the southern world (pictured) was tracking a strong radio source close to the full moon. John Bolton was now director of this instrument, with its 210-foot dish antenna poised on great steel trunions above a concrete tower. After the debut of discrete radio sources in Sydney, Bolton had established the Owens Valley radio observatory, in Pasedena, USA, and had there been puzzled by obscure objects which flung out enormous power in radio emission — among them the original 3C48 and another listed as 3C273 (pictured).

Now — in August 1962 this latter radio source was in a position in the southern sky where the Moon would ride by and block its radio waves from reaching Earth. A group of observers, led by Dr. Cyril Hazard, monitored the so-called "occultation". The position of the 3C273 object in the sky — which was believed to be two closely related radio sources — was highly important to obtain optical indentification. Since the position of the Moon at any moment is known with great accuracy, Hazard and his colleagues used it to obtain the exact point in the heavens from which the copious radio energy emanated.

That area of the sky is lush with a congregation of Milky Way stars; and there was then no optical telescope in the southern world capable of resolving the object to the degree where answers could be found to the questions which arose on the nature of such a small star-like object. The details of the occultation study were sent to Dr. Maarten Schmidt, a brilliant Dutch astronomer working at Mount Palomar. Interest there had already been aroused by the mystery of 3C48 and similar known objects.

A powerful telescope was used to track down the object from the exact co-ordinates sent from Australia. Schmidt found a small star — no different in appearance to a million such stars in the Galaxy. Dr. Schmidt broke down the light spectrum of this faint, fuzzy object — and the Red Shift of its light showed it to be racing away at a recessional speed of 50,000 kilometres each second.

The laws framed on the expanding universe concept said the small, shining object was some 2,000-million light years distant (pictured). To be seen as bright as it was from that very remote distance, its luminosity would have to be between 40 and 100 times brighter than a normal galaxy. Man had no rules on the physical behaviour of matter to explain that vast output of light and energy from this so-called star.

Astronomers named it a "quasi-stellar radio source" — but that soon came down to the term which has come to represent one of the great enigmas of the cosmos — the quasar.

Not all quasars emit powerful radio signals, but all are reckoned to be highly compact with the material of 100 normal galaxies crushed into an area similar to that filled by the Solar System. They line the rim of the observable universe, shining back across the cosmos from as far as we shall ever see.

For all that time — for 10,000-million years, at least, light has been travelling toward Earth . . . and Maarten Schmidt calculates that, in all, we can now see the fires of 14-million quasars burning in the far cosmos. The light has travelled for 10,000-million years and, thus, shows a picture of the situation which existed then. In real time, however, most of the original 14-million galaxies (created when the cosmos was young) are probably no more. Schmidt believes that, at this point in our history, the actual number of quasars in existence is no more than 35,000. There can be no certainty about this since proof will not reach us for many more thousands of millions of years. What is certain is that the quasars — as are the

Far beyond the starfields of the Centaurus constellation from 80 million light years distant, tens of galaxies glow through the beams of the suns of our home galaxy. Of different types, some bright some dim, they are all separate island universes. Among them are strong emitters of radio signals and the strongest is the brightest, near the top of this picture, NGC 4696. This photograph required 45 minutes exposure. (Photo: Dr. R. R. Shobbrook.)

globular clusters in our own Galaxy — are a feature of the earliest period of cosmic creation, that they were formed soon after the greater universe was formed.

We cannot know their fate, as we do not yet know their true nature. Whether they represent the *beginning* of evolution of galaxies . . . or their *ending*, is not yet known. No satisfactory explanation has been found for their enormous power. And so baffling are these enigmatic objects, some senior astronomers are looking again at the behaviour of light. They are asking — Is the Red Shift from quasars an illusion and not a sign of true distance?

There is no ready answer. The mystery uncovered by the Australian lunar occultation of 3C273 is a beautiful puzzle which exploration of the invisible universe has produced. And there are others.

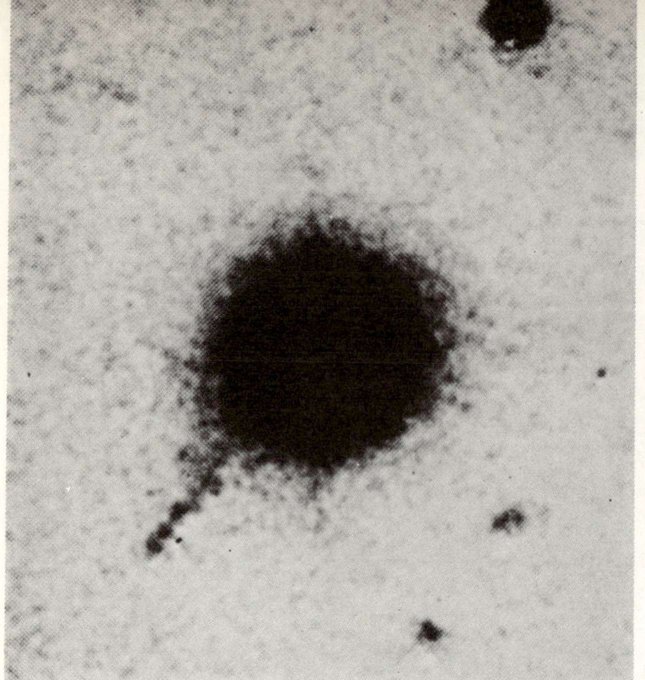

First of the Quasars. This is the optical view — greatly enlarged — of the first identified quasar, listed as 3C273. Its position was determined by Australian radio-astronomers at the CSIRO observatory at Parkes, New South Wales. The light from this remote object is brighter than 40 normal galaxies and has to travel for 2,000-million years to reach Earth. Its discovery rolled back the horizons of the cosmos for mankind to a time and distance measured beyond 10,000-million light years. *(C.S.I.R.O. photo)*

One of the most distant quasars, discovered at Parkes in 1966, has a speed of recession of 91 per cent velocity of light and therefore appears to be of the order of 6000 million light years away. This means that the light which we see now from this object started travelling towards us before the Sun, Earth and other planets came into being. This object (marked by indicators) is also the most luminous single object in the Universe, pouring out about 100 times more energy than all the stars in our galaxy — The Milky Way. *C.S.I.R.O. photo*

Midget stars with astonishing power

Many discoveries have been made in the invisible universe. Sources have been found of ultraviolet, gamma and infrared rays, of great gouts of x-rays emanating from the fringes of galaxies, from the hearts of strange star systems, from quasars and star-groups — some from blank spaces in the universe — and mostly these have been located by equipment flown by rockets, or balloons, or on orbiting spacecraft.

There has been none so bizarre, so appealingly mysterious as the pulsating dim midgets now believed to be the re-incarnation of giant suns that exploded in the past. These were first revealed by radio waves detected on Earth.

The simple start to this story is that radio-astronomers in Cambridge, England, swept the sky in 1967, searching in wavelengths shorter than the normal radiation for celestial emitters; and they made the surprising discovery of a beeping radio signal which was so consistently accurate some people came at once to the thought that it was the product of some intelligence — perhaps a radio navigation beam across the Galaxy for travellers from another civilisation.

In terms of astronomy the true facts proved to be even stranger. These pulses of radio signals were as regular as the best atomic clocks.

Excited astronomers gave them the prefix of LGM — for Little Green Men. And while that was fantasy, the truth later emerged as even more fantastic.

So strange was this phenomenon, radio-astronomers across the world at once scoured the heavens for similar signals — and found them. Australian workers, with an array of radio-telescopes and instruments found 30 such objects in the southern sky, of the first 40 discovered. And one of them was in the heart of the supernova, the Crab Nebula.

Ideas and theories on the nature of these invisible pulsing objects crystallized from world-wide study. The pulsing source in the Crab proved to be strong confirmation of what is now accepted as the explanation for the objects that have come to be called — Pulsars.

The year of 1968 is the Year of the Pulsar in the history of exploration of the cosmos. In that year, pulsars produced the under-

standing that a star somewhat bigger than our Sun can explode at the end of its radiant life and can, in the utter devastation of its eruption, create a new, small, tiny, dark star — a star formed entirely of the sub-atomic particles called neutrons (which, incidentally, were first discovered in 1932, in the same university city which found the pulsars — Cambridge).

A pulsar, then, is a neutron star. The strangest of all known stars, it is composed of the most dense material in the universe and is created by the "imploding" force of an exploding sun.

A giant star may shrink down as a white dwarf to about the same size as Earth . . . but, a pulsar measures no more than a few kilometres in diameter. Its power output is another matter entirely.

More than a hundred pulsars are now known to be flashing their signals across the Galaxy — and some of them are shown to be point sources of copious flows of x-rays; particularly the pulsar in the Crab Nebula.

The questions were asked — How does a midget star, formed of neutrons, spit out rapid radio signals like machine-gun bullets? How does a tiny star like the pulsar in the Crab Nebula fling off x-rays with energy 10,000 times greater than the Sun's total output?

The answers were quickly found. Pulsars rotate at phenomenal speed. The tiny pulsar in the Crab (pictured) is among the most rapid of known pulsars. It expels its radio signals 1,800 times each minute, because it spins at a rate of 30 times a second. Astrophysicists argue that its dense nuclear structure, its fierce gravity, the strain on its surface (which causes miniature star-quakes) all contribute to the burst of energy flung out . . . in wavelengths covering the whole spectrum, including a small showing of visible light. The extreme output of x-rays probably comes from material striking against the bizarre surface of this strange star.

Pulsars are now known to slow down, by about one part in 2,000 each year. Thus, a brand-new pulsar would spin even faster than the one in the Crab; older ones are much slower in rotation. But, though they are slowing down, these re-incarnated stars are capable of pulsing out enormous energy for many, many millions of years.

THE COSMOS IS YIELDING ITS SECRETS

Galileo Galilei first saw the moons of Jupiter on the night of January 7, 1610. He wrote how he was "led by some fatality" to this event in the exploration of the universe. It also led him, fatally, into the hands of the Inquisition, to recant, under threat of burning at the stake, his belief that Earth rotated around the Sun. Whilst under close arrest, he died, blind, alone, but undefeated, in 1641.

Truth was not choked by this crime; telescopes were not blindfolded. Man has come a long — and humbling — path to the giant instruments now used to unravel the riddle of the cosmos. Today, astronomers, cosmologists, philosophers discuss cosmic origins without fear of dogmatic retribution.

The centuries since Galileo have carried human thought to the threshold where, this century, a Belgian priest and astronomer, Georges Lemaître, could propose a theory for total creation and say —

"The evolution of the world can be likened to a display of fireworks just ended . . . some few red wisps, smoke and ashes. Standing on a well-chilled cinder, we see the slow fading of the suns and try to recall the vanished brilliance of the origin of the worlds."

The journey in this book has been made outward from the Sun and its Family of Planets, through the stars of the southern sky, across the Southern Milky Way, and beyond the nearest galaxies to the rim of the visible universe. It has shown how the slow fading of suns is now understood — when two decades ago man did not know how to age-date the stars. We have caught glimpses of the 'vanished brilliance of the origin of the worlds', and have learned that light from the far flung quasars bridges enormous time across a cosmos man did not know existed until half-a-century ago. Travelling, in our minds, across that ocean of 'island universes', we have seen the stages of cosmic evolution.

Where, now, do we stand?

We are very near to the final clues from which astronomers can build the ultimate explanation of creation. Almost certainly this will be a 'model' of a cosmos that commenced with the Big Bang of an exploding, primordial fireball. This is the true Genesis in which, cosmologists now deduce, all matter was constructed. Pure energy created atoms, atoms went into molecules — and gas into galaxies. And when did this happen?

All the clocks of the cosmos tick in unison, sounding back unerringly to the same point of past time. The light from the furthermost quasars speaks of a epoch in the young cosmos; the aged globular clusters are of that same generation: the elements show a dating that is confirmation, and the chemistry of the universe — being no different in any way than that which forms the crust of our Earth — tells the same tale of age.

The shells of fleeing galaxies, ranging out through space and time, give similar indication. The hands on all the cosmic clocks point back to the creative event, to 10,000-million years ago. Despite this the idea of a Big Bang start for the cosmos has been challenged. It was a counter claim which argued a steady-state universe in which new matter was constantly created to compensate for the outward rush of the galaxies — that 10 billion years ago, or forward, the universe would still look the same as it does today. Few astronomers, however, find room for that theory after the explosion of knowledge and technology which has come in the decades since man left the cradle of Earth and opened the Space Age.

Such a stupendous eruption as envisaged in the Big Bang would, predictably, create low-key background radiation of a certain intensity, and cause a diffuse flood of x-rays across the universe. These have now been found. As

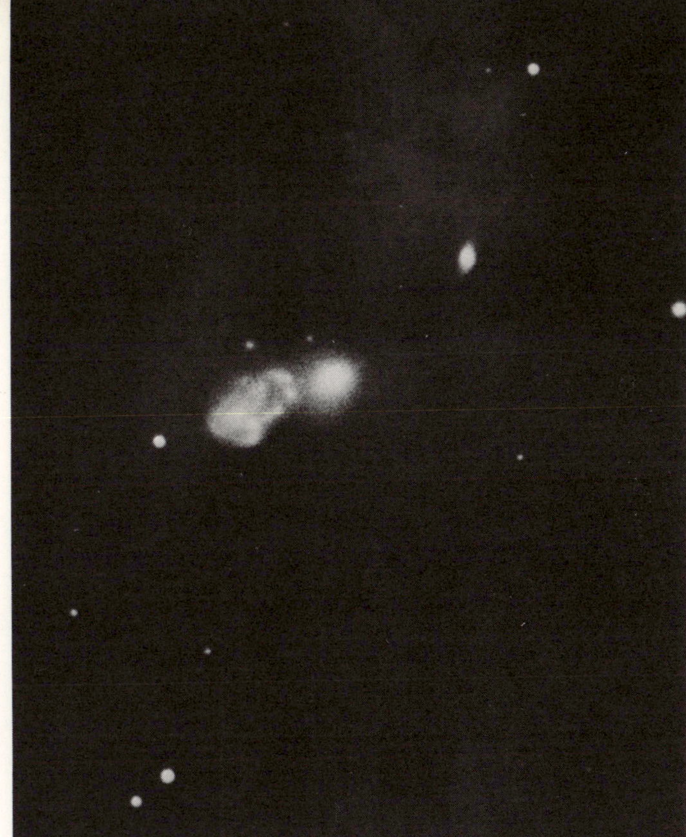

Four separate and very distant objects in the southern sky, not seen like this before, present baffling features which astronomers now seek to explain. Like tiny marine creatures in these four separate photographs, these are distant southern galaxies ranging out to 100-million light years.

Glowing gas smudges their outlines and in one case forms a smoke ring in the heavens. Hazy against the stars of the Milky Way (in the foreground) these remote galaxies are believed to be moving bodies in collision with massive clouds of gas and dust far off in inter-galactic space.

Experts in this field believe that in such circumstances the colliding gas cloud leaves the star bodies and clusters untouched but drags out the galaxy's gas clouds with considerable force, spreading tails of star-building materials like wispy antennae across hundreds of light years.

(K. Freeman, AAT, Siding Spring)

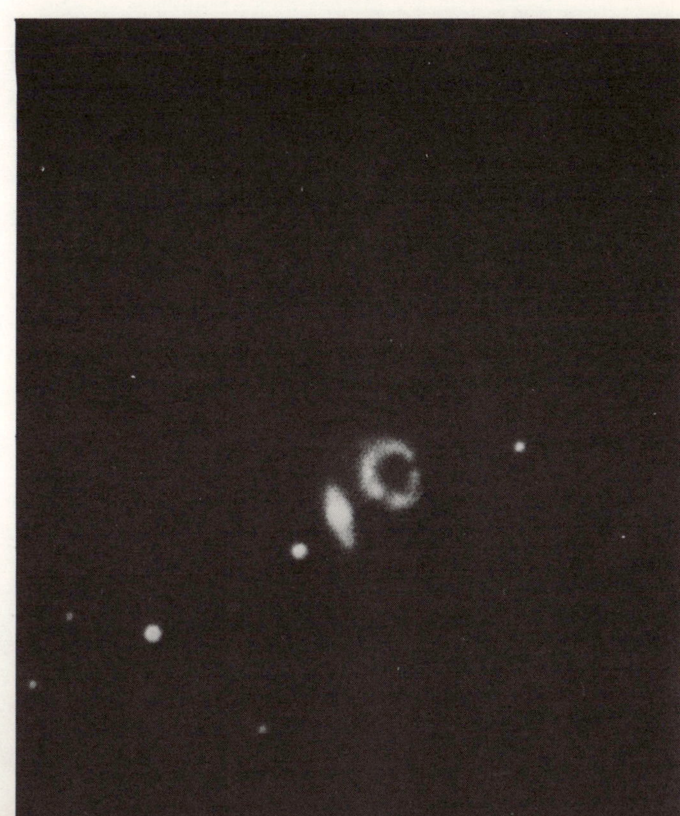

well, there is no sign of constant creation in the preponderant, brilliant, elliptical galaxies which dominate most clusters of flying star-worlds. Seek as they may, astronomers have not found one young supergiant star in these sterile galaxies.

Always the signs point back to the explosive genesis of 10-billion years ago. Only a few final factors remain for precise and accurate measurement, and key clues to some of these are in the southern sky.

In the search for conclusive parameters, two factors dominate the choice of the final 'model' we shall accept for original creation. And, in these, the behaviour and nature of the nearest galaxies — The Clouds of Magellan — and also of the local cluster of galaxies, are as important as those objects on the rim of the visible universe. The remote worlds tell a tale of long ago; the nearest galaxies are the most recent cosmic chapters.

The two key factors now being keenly sought, 40 years after Hubble unfolded the expanding universe, are: (a) the increased rate at which the galaxies recede for each million light years of distance; and (b) any rate of slow-down, or deceleration, in the outward flight.

The first factor (a) is known as H — for the Hubble constant; the second (b) is referred to as Q . . . and it is certainly a key question. Against Hubble's constant, when it is precisely established, Q will confirm a theory that all matter thrown out in the original Big Bang is being slowed down by gravity, by the pull of the total universe itself. Will the galaxies eventually lose outward power? Will they return to the point of origin to another doomsday event and subsequent re-creation? Put more simply, are we living in the oscillating universe of the kind the Mayan astronomers envisioned?

The Southern Universe has a key role in this puzzle. The eminent deep-space explorer, Dr Allan Sandage, is a world leader in this field and says that in-depth exploration of the southern sky is crucial to accurate definition both of the Q factor and the Hubble constant. "The big southern telescopes are opening a new frontier, because, only in the southern sky can we find such a pure outward Hubble flow of galaxies. The 40 or 50 galaxies in the south polar region of the sky hold the key to the time-scale for the rate of expansion we can apply to the whole universe." As well, in his view, the Clouds of Magellan are the Rosetta Stone of creation. He says, "They will provide many surprises in the next few years".

The Commissioning Astronomer to the big Anglo-Australian Telescope — the first working major telescope to open in-depth exploration of these critical regions in the south — is Professor S. C. B. Gascoigne. "The southern sky now holds tremendous interest and importance in world astronomy in general. And the work in the south is only just opening out. There is vast promise in the

The Sombrero — This unusually bright galaxy, catalogue number NGC 4594, is found in the Virgo constellation, but is of southern interest in its similarity to the remarkable Centaurus galaxy.

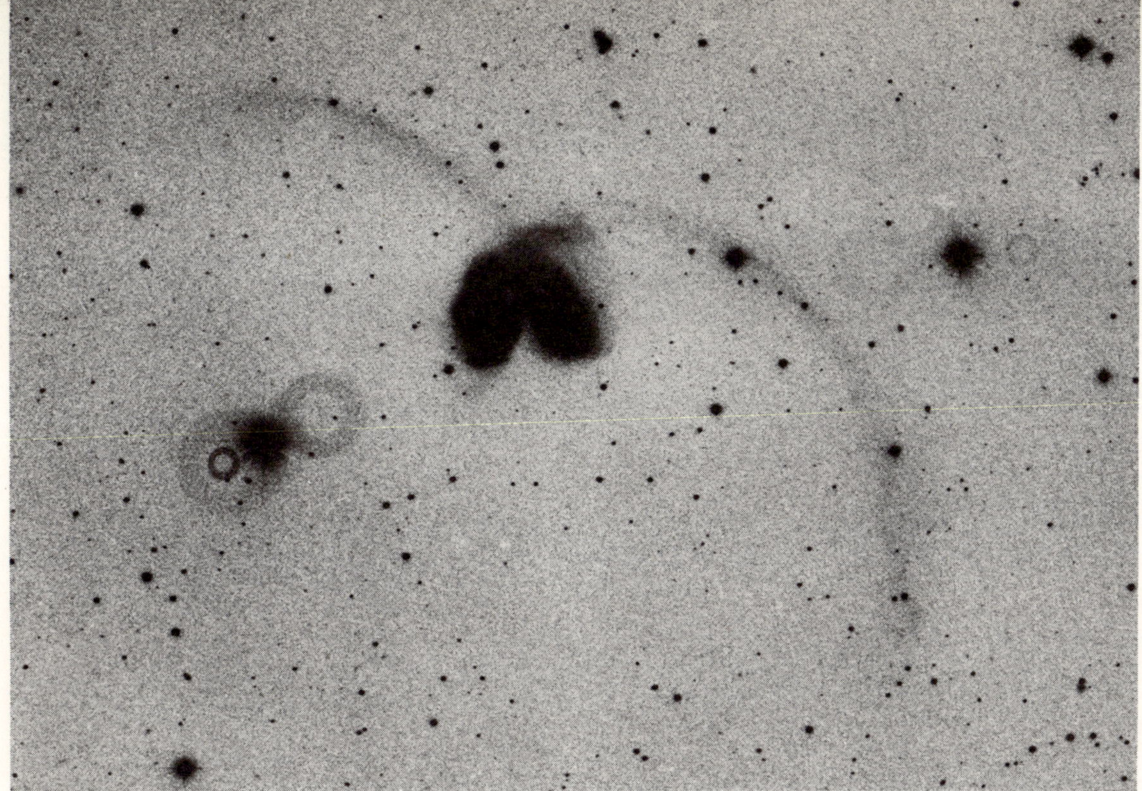

The Antennae Galaxies — This bizarre object is an unexplained event in the universe of galaxies. Two star systems, locked in embrace and trailing plumes of gas across space like the whiskers of insects, their shape is now believed to be due to tides of gravitation between colliding systems.

The "Antennae" galaxies are in the southern system of Corvus — found northward from the upright Southern Cross. They are about 50 million light years from Earth and among the 1,000 nearest external galaxies. Such photos — printed in negative rather than conventional white on black — show up finer detail of the tenuous gas tails. Pictures like this require hours of exposure. It was taken by Dr. Ken Freeman at Siding Spring Mountain with the 3.9-metre telescope. Pointed objects in the foreground are stars in our own galaxy while the rings are light abberations in the powerful telescope.

nearest galaxies, in southern members of the local cluster of galaxies, and in the abundance of matter between them and between the stars".

The 3.9-metre instrument, and similar telescopes being prepared elsewhere in the southern hemisphere will rake the southern heavens in the coming years. "There is an age ahead that holds rich promise of discovery", says Professor Olin Eggen. Its promise arises from the revolution which technology has brought to current astronomy.

The virgin reaches of southern sky offer scintillas of starfire to which previous telescopes have been blind. The AAT can gather photons, particles of light from single atoms that were in remote quasars which despatched their radiance across the great gulf of space 10,000-million years ago when they were then already receding at 80 per cent the speed of light . . . *at 240,000 kilometres a second.*

Every single photon collected on the superb mirror of the AAT is, thus, precious to building the light picture of the face of the cosmos as it was some 10-billion years ago. Just as important is the fine discernment the magnificent instrument makes possible within the structure of the Magellanic Clouds, and which touches on the origin and nature of life itself. The examination of exploded stars in the Large Cloud of Magellan by Dr. Don Mathewson, of Mount Stromlo, is now unparalleled. The AAT has shown him fourteen examples of supernovae *in an external galaxy* with detail not seen before. One of these exploded stars was a mere 200 years old when the light of its destruction started its journey from the Cloud to Earth.

What is there to learn from this? The nature and velocity of the material thrown out are keys to how the elements of life are distributed in our own Milky Way Galaxy. Stars make heavier elements both in their living and in their destruction. Strewn by explosion across space (pictured) in the Large Cloud and in the Milky Way, they enrich younger stars with elements ranging up through the metals. And the metal-rich stars can form planets with the elements which support life on Earth. "The catalysts of organic life are formed in such stars," says Dr. Mathewson.

To such fascinating aspects of astronomy technology has added new powers of distinction, and detection. The AAT instrument, backed by the finest possible array of technical and electronic equipment, can record a ratio of photons from distant objects twenty times greater than was possible with the world's finest telescopes two decades ago. So steep has been the increase in the collecting ability and the storage and recording of photons of light, that many astronomers believe the ultimate in photon collection and retention has been achieved. That is why Professor Eggen is able to say — "We are now seeing as far as we shall ever see."

From the Siding Spring Observatory, astronomers now not only see further than ever before — they see a great deal more. Working side-by-side with the big AAT is the British-owned wide-angle Schmidt telescope. This camera telescope is the largest of its type in the southern hemisphere and as big as any in the world. The 1.2-metre mirror is being used to collate the first detailed photographic survey of the southern sky. This current undertaking by British astronomers, led by Dr Russell Cannon, of Edinburgh, is in conjunction with European observers using a smaller, similar Schmidt from the Cerro Tololo Observatory in Chile. The Schmidt on Siding Spring Mountain is photographing the Southern Universe — in depth and in detail — with blue-light sensitive plates. The European-owned Schmidt (1-metre) telescope in Chile is photographing the same areas in red light.

The Siding Spring Schmidt has already produced scenes of beauty never before photographed in detail (pictured). The value of this work, however — tremendously exciting as it is to the astronomers probing into virtually unexplored corners of the Southern Universe — will have added value when matched with surveys made from Australia of radio sources.

What new wonders will all this combined research reveal?

"No man can say," comments Professor Olin Eggen. "Who could have foretold the quasars, the going back over 10-billion years of time, the pulsars, or the x-ray sources? What is certain is that the coming years are going to bring discovery as exciting as any yet made."

There is great promise, also, in space astronomy. In the next decade the first large optical telescope will go into orbit above Earth's atmosphere. It has already been designed and building has started. The U.S. space agency (NASA) plans this instrument to be about 60 feet long, 12 feet wide, with a 120-inch diameter reflector mirror. The launching of this giant orbital telescope will be part of the so-called Shuttle Program. The shuttle service will put the LST (large space telescope) about 450 kilometres above the planet's surface, and, from there, a new age in the study of the universe might well open.

The LST will not be manned. So delicately precise will its focussing have to be, the presence of a human heart in the weightless conditions of orbit would cause perturbation of the instrument's built-in guidance system which will hold the focus of the telescope to an accuracy of 1/5000th of a second of arc; and that is the width of a human hair seen from two miles away. The telescope is planned to look into the Clouds of Magellan, to study the central regions of the Galaxy, and to advance general study into the origin of the universe.

There has been no period in astronomy to match the one in which we now live. There has been no comparable explosion in knowledge or in the ability to explore. The prime burning issue in science: the beginning of the universe, the origins of man, that supreme intellectual quest is much closer to elucidation than most people realise.

The outward thrust of the galaxies, the revealing red shift in their light, the strange behaviour of matter in the newly-found bizarre objects, the abnormally young star clusters in the Clouds of Magellan, the prospect of looking into chemical abundances in the hidden heart of the Southern Milky Way, the widening band of life-bearing molecules in the Sagittarius clouds, the heavy elements thrown out by the Magellanic supernovae — all are grains of knowledge building toward a complete picture.

Certain people may question whether this effort and expense is worthwhile in practical terms. But who can write a price ticket for knowledge? Astronomy today stands at its widest frontier with all the signs of impending discovery such as attended other sciences in the past; x-ray diagnosis came from ionized gas research, the conquest of disease came from pure medical science, transistors from solid state physics, and dozens of today's common facilities rose out of science for which few men could see a practical purpose. Michael Faraday replied when asked the practical use of the discovery which made the generation of electric power possible — "What use is a new-born baby?"

Astronomy is not a new-born baby; it is the most ancient among sciences and the dividends it has brought over 5,000 years are incalculable in terms of the power of thought.

No one knows what benefits will accrue from modern astronomy — other than knowledge — or even if any benefits will ensue; but, if history is a guide, today's research will be valuable to future generations; for, in the pure spirit of enquiry, it is part of the human story.

The courtyards of the cosmos may be surveyed by automoton telescopes in space, but, men will make the new discoveries from Earth in what is left of the 20th Century.

In this great adventure the Southern Universe may hold the vital clue, or produce the one spectrum which will enable astronomers to finally write a birth certificate for the great universe.

A globular cluster, known as NGC2808, is at a distance from earth of 23,000 light years.
(Photo: Vince Ford, Mount Stromlo — 40-inch)

GLOSSARY

Absorption Lines: see *Emission Lines.*
Asteroids: rocky rubble distributed in the belt between Mars and Jupiter, ranging from pebble-size to the largest asteroid, Ceres, some 700 kilometres diameter.
Alpha: prefix used for brightest star in the constellations; i.e. — Alpha Centauri.
Binary systems: star couplet or two stars in common orbit, which are more the norm than singles.
Blackbody radiation: predicted low-intensity radiation which fits with "Big Bang" theories and of which some evidence has been found.
Black Holes: believed to be "sinks" of fierce gravity caused by imploded suns from which no light can escape.
Cepheids: variable stars with periods of fluctuation in their light that is linked to their intrinsic brightness; thus they can be used as "space candles" for standard distance calibration.
Corona: tenuous outer atmosphere of Sun, seen most clearly at times of full eclipse by the Moon.
Chromosphere: solar layer below corona, believed to be about 16,000 km deep.
Clouds of Magellan: name given to two nearest galaxies to Milky Way by early travellers to whom they seemed like two white fuzzy patches circling the South Celestial Pole.
Comets: itinerant fragments of gas and dust caught in far-ranging orbit round the Sun; believed to number in the millions.
Cosmos: from the Greek *kosmos* and a term embracing all creation — the 'island galaxies' and all matter that might exist in between.
Cosmology: study of the origin and nature of the Cosmos.
Cosmogony: term now usually reserved for study of the Solar System.
Deuterium: name given to a form of hydrogen formed in the fusion inside stars from simple hydrogen to which an additional particle has been added. Deuterium atoms bonded into the normal H_20 of water molecules form 'heavy water', the rare substance used in nuclear reactors.
Dwarf stars: usually white dwarfs; i.e. degenerate stars which have shrunken in the final processes of stellar decay.
Ellipticals: egg-shaped galaxies, mostly massive and in the majority in far-ranging clusters of galaxies; they commonly show an absence of star-making material and a total lack of young, very hot, supergiant stars of the type common in spirals.
Emission lines: excited atoms emit radiation which carries lines characteristic of the element involved; usually known as 'fingerprints.' Matter between the star and Earth can also radiate from absorbing such radiation; this also carries lines known as *absorption lines.*
Expanding universe: term given to phenomenon noted by Edwin Hubble — now commonly accepted — that the galaxies are receding in all directions at speeds that increase with distance. This paradox is often likened to a boiled pudding in which raisins move apart as the pudding is cooked. See — *Hubble constant.*
Extra-galactic: term to refer to objects or events outside the boundaries of the Milky Way.
Fixed stars: term used mostly in navigation for stars that are so distant they appear never to change their position relative to Sun and Earth; literally, there are no fixed stars since all suns, clusters, and gas clouds of the Milky Way revolve at different speeds round the Hub of the Galaxy.
Fusion: term used in nuclear physics for the union between atoms of the light elements, such as the fusion of simple hydrogen in stars which can proceed on to the final creation of heavier elements as the number of atomic particles is increased. The process of fusion

liberates matter in the form of energy, such as the Sun or man-made H-bombs; as against fission in the heavier elements — uranium, thorium, plutonium — which also releases enormous energy.

Galaxy: world or universe of stars and gas which occupy the cosmos in all directions and number many billions.

Galactic cluster: close-packed colonies of suns aligned along the gas disk of the Milky Way: see *Globular Clusters*.

Galactic year: the Milky Way Galaxy turns on its Hub in space. The Sun and planets take about 250-million Earth years to complete a single turn about the heart of the Galaxy; that is known as a galactic year (some astronomers call it a 'cosmic year'). Not all stars move round the Galaxy at the same speed.

Gamma rays: high-energy, intensely short, waves of radiation in the far end of the electromagnetic spectrum of radiant energy which is released by disturbance in the structure of atoms.

Globular clusters: tight-packed colonies of stars found in the outer "halo" regions of the Milky Way and representing the oldest known creation in the cosmos. Clusters can be seen in nearby galaxies but these are usually much younger. Assessed ages of globular clusters is around 10,000-million years.

"Halo" — Milky Way: outer regions without prominent gas and populated by aged stars; believed to show the true shape of the early Milky Way Galaxy — before the main material fell back to form the present disk.

Hubble constant: concept of the expanding universe says the speed of recession of galaxies increases progressively with each million light years of distance. The Hubble constant is thus a mathematical configuration closely linked to the true size and the speed of the expanding universe.

Infra-red — short-wave invisible radiation near the visible light segment of the electromagnetic spectrum; can be recorded on red-sensitive photographic plates and since, like radio wavelengths, infra-red can find its path through dust and gas clouds it allows deeper penetration of normally obscure areas.

Interstellar medium: gas and dust forming material between the stars.

'Invisible' universe: refers to energy and activity in the universe which cannot be seen in normal light but is detectable by other means; i.e. radio, x-ray detection tubes, red-light sensitive plates.

Ionosphere: part of Earth's upper atmosphere absorbing harmful energies from space, which reflects short-wave radio back to Earth.

Ionized gas: matter in which radiation has separated electrons from atoms; i.e. — plasma.

Light year: a measurement of distance — how far light travels in a year; i.e. 9.5-million-million km.

Metal-rich: term for obviously younger stars formed from gas and dust clouds bearing more complex atoms fused and fabricated by the processes of earlier stars and probably distributed through space by unstable stars that have exploded: see supernovae. Old stars are usually metal-poor.

Milky Way: commonly used for the white band arching the night sky. Visible from everywhere on Earth because the Sun is buried inside this disk of stars and gas. In this book Milky Way refers to the galaxy, not just to the white band seen across the sky.

Magellan: see Clouds of: named for Ferdinand Magellan the explorer of the Pacific. Two external galaxies of immense value to astronomy.

Nebulae: refers to clouds of gas and dust usually associated with star formation; sometimes known as planetary nebula when shrugged off in circular or ring formation from emerging stars.

Neutron stars: believed to be created by the imploding forces of exploding giant suns; in which gravitation is so fierce atoms are crushed and singular immensely dense bodies are formed from neutrons. These revolve at fantastically high speeds and emit radio and x-ray and very small amounts of light. Theory says these objects can create "black holes", or gravitation sinks. Usually known as pulsars.

Observable universe: term to describe limit imposed by the nature of light which we shall never surmount. The known rim of the total universe reveals faint objects receding at close to the speed of light. Beyond that, light cannot escape to travel back to reveal its existence.

Occultation (lunar): using the known precise position of the Moon to locate a faint object in space. The technique is also often applied to stars which occult each other and to larger and more distant objects.

Parallax: method of using angles of sight from the position of Earth each side of the Sun to obtain distance to stars.

Parsec: astronomer's yardstick; one second of arc each side of Sun using parallax (see above) gives a point where the two lines meet in space: i.e. 3.2 light years — a parsec. Can be used as kiloparsec (kpc) or even as megaparsec (mpc).

Photosphere: turbulent layer of the Sun's structure which extends some 130,000 kilometres from the surface inwards.

Plasma: state of ionized gas: see *ionized gas.*

Pulsar: a dark, dense, intensely small and powerful star created from neutrons in star explosions — see **neutron stars.**

"Q": letter used as symbol to indicate any slow-down in the rate of recession of galaxies; part of the concept of the oscillating universe. See *Hubble constant,* and *Red Shift.*

Quasar: shortened version of quasi-stellar-radio-source; i.e. very distant objects looking like stars but separate island galaxies and radiating amounts of energy by processes not yet explained by any known laws of physics.

Radio astronomy: detection of objects and study of their nature by characteristics of their radio wave emissions.

Red Giant: state of a star that has consumed most of its hydrogen and has expanded to many times its former volume — as a law of stellar dynamics. The Sun will become a Red Giant in about 4,000-million years.

Red Shift: feature of light spectra from distant rapidly receding objects is the shift toward the red end of the spectrum of the lines of known elements; thus the extent of the shift to the red (caused by speed stretching out the length of the red light waves) indicates speed of recession which in turn reveals the approximate distance of the object. See, *Hubble constant* or *"Q".*

Reflector telescope: instrument which catches the light from space on a reflecting mirror so that it can be bent back along several paths for use by astronomers, such as the examination of the spectrum in a spectroscope, for photography, visual observation, or specialist study. As opposed to the earlier-type refractors which make use of lens set apart by focal distances, as in binoculars.

Ultraviolet: part of the shorter end of the electromagnetic spectrum; of longer wavelength than gamma and x-rays but still highly dangerous to forms of organic life.

Sagittarius: constellation set against the denser clouds of the Milky Way's disk and important since this area marks the central Hub of the galaxy.

Schmidt telescope: special wide-angle telescope which allows deep photographic penetration of wider areas of sky than larger instruments; the British 48-inch at Siding Spring in Australia is a superb example, the biggest in the south and one of the largest in the world. Some photographs taken with this fine instrument are displayed in this book.

Solar wind: outflow of the Sun's atmosphere impelled by the fusion processes in the solar interior.

Space "candles": see *Cepheids.*

Speed of light: (c) 30,000 km/p/second; see *Light year.*

Spirals: see *galaxies.*

Sun spots: dark pockmarks on the Sun's face caused, it is thought, by spiralling fields of magnetic forces from inside the solar structure; they are darker because they are somewhat cooler; and they give off radiant energy which bombards Earth's upper atmosphere and cause disruption to normal short-wave radio.

Supergiants: stars many times greater than the Sun; usually blue-white-green, very hot stars consuming their fuels at prodigious rates; thus, stars which live relatively short lives.

Supernova: term used for exploded star. Supernovae are thought to occur in a stage of evolution of unstable stars where the inward pressure of gravity is exceeded by the outward thrust of internal processes. These objects assume high interest because they may fashion the catalysts of life and of our type of world and scatter these elements through space. Supernovae also create situations leading to pulsars and black holes.

Transit of Venus: passage of the planet across the face of the Sun; used as a yardstick to measure the solar distance by Captain Cook's first Pacific expedition and by subsequent astronomical observatories in Australia. The Cook measurement of the transit of Venus was compared with a similar one taken on the other side of the world and the angles were extremely accurate in providing the required measurement. See *parallax.*

White dwarf: see *dwarf stars.*

X-rays: part of the electromagnetic spectrum; one of the shorter wavelengths of radiation now forming a separate part of astronomy since their presence in space is revealing surprising data on invisible objects such as pulsars, and black holes, as well as the mysterious reactions which emit power from the hearts of certain types of galaxies.